STYLISH
STORES II

时尚专卖店 II

VOL. 1

深圳市艺力文化发展有限公司 编

华南理工大学出版社
SOUTH CHINA UNIVERSITY OF TECHNOLOGY PRESS

·广州·

图书在版编目（CIP）数据

时尚专卖店 II = Stylish stores II ：英文 / 深圳市艺力文化发展有限公司

编 . — 广州 ： 华南理工大学出版社 , 2014.3

ISBN 978-7-5623-4176-5

Ⅰ . ①时… Ⅱ . ①深… Ⅲ . ①商店－室内装饰设计－作品集－世界－英文 Ⅳ . ① TU247.9

中国版本图书馆 CIP 数据核字（2014）第 038111 号

时尚专卖店 Ⅱ　　Stylish stores Ⅱ

深圳市艺力文化发展有限公司　编

出 版 人 ： 韩中伟

出版发行 ： 华南理工大学出版社

（广州五山华南理工大学 17 号楼，邮编 510640 ）

http://www.scutpress.com.cn　　E-mail: scutc13@scut.edu.cn

营销部电话： 020-87113487　87111048 （传真）

策划编辑 ： 赖淑华

责任编辑 ： 王岩

印 刷 者 ： 深圳市新视线印务有限公司

开　　本 ： 635mm×1020mm 1/8　　印张 ： 55

成品尺寸 ： 245mm×300mm

版　　次 ： 2014 年 3 月第 1 版　　2014 年 3 月第 1 次印刷

定　　价 ： 598.00 元（共 2 册）

PREFACE

Nowadays, a store is more than just a new address for purchasing products. New stores are simultaneously new event areas, new performance stages, sometimes even new landmarks of a city. They are stylish and sophisticated, equipped with an elaborate design concept.

The role of stores has changed in the past few decades. The transformation is a continuing process shaped by various trends, shopping habits and technological developments in retail. Stores have become statements, celebrating the brand and its products. They tell stories and communicate the philosophy, values and responsibilities of a company. At the same time, they are spots for happenings of various kinds bringing like-minded people together.

Stores are "real" and "physical" places creating an engaging shopping experience. They bring a unique atmosphere into being and evoke emotions. Their design creates a unique "world" – the world of a brand – and tells a story of this particular "world". In this way stores offer multiple chances to engage with the brand and its values.

One of the greatest influences on the transformation in retail environments in recent years is the technological process. It has changed our shopping habits as well as the shopping experience and store design. A liaison of the "virtual" and "non-virtual" is becoming more prevailing in store concepts, thus shaping the overall character and appearance of shops. iPads and interactive screens are just as important in store design as display tables and shelves.

Once these transformations are captured in a book like "Stylish Stores" the story is being told… written and sealed… For the here and now it is a wonderful source of inspiration, for the future a formative retrospect. By selecting a variety of projects originating from all over the world "Stylish Stores" tells different stories from a certain period of time.

Enjoy your journey!

plajer & franz studio

CONTENTS

FASHION STORES

FASHION STORES

Who's Who Store

Designer:
Fabio Novembre

Design Team:
Dino Cicchetti, Giulio Vescovi

Client:
Max-Company

Location:
Corso Venezia, Milano, Italy

Area:
85 m²

Photography:
Pasquale Formisano

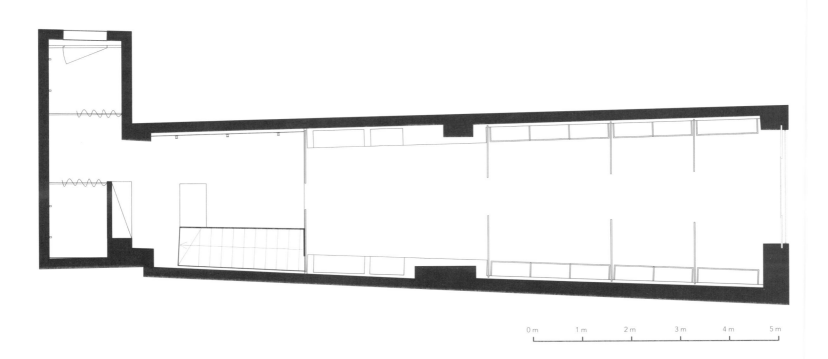

0 m   1 m   2 m   3 m   4 m   5 m

Who's Who, a symbol of modernity and feminity, has confirmed its international vocation and presented the new concept of charactised boutiques.

In order to develop the new retail strategy and create an innovative design, Who's Who engaged Fabio Novembre, an eclectic, imaginative and contemporary architect who, with a surprising project, succeeded in highlighting the company's DNA, with a strong, meaningful and long-sighted concept. The boutiques become an ideal set for a meeting between a man and a woman, represented by outsized sculpted glass figures, that seem to be walking slowly across space while their hands search for each other, until they brush against each other, in a free interpretation that remind one of Michelangelo. A surreal scene, in which the wrought steel walls reflect a multiply reality, making it fluid; the floor, slowly sloping towards the side walls, is the only hint showing the direction to go, because, as the architect points out, "only the spark of love can light the flame of creation", of any kind. "We like to say that the name Who's Who refers to the search for an identity, to the constant leaning towards something that is not the self", Mr Novembre insists.

The ideal location for the launch of this ambitious plan is Milan's "Quadrilatero della Moda": the first shop opened recently in Corso Venezia 8.

This important retail project was strongly supported by Massimiliano Dossi, the head of the company, and the opening of another single-brand boutique, this time in Forte Dei Marmi is scheduled for the end of March, Forte Dei Marmi is a strategic Italian location, given its status of internationally renowned sea town.

In 2013 the company scheduled the opening of at least four other boutiques in areas in which the brand has been quite successful in the past several years: China – Middle East – Russia. Finally, the three-year plan, which shall see its end in 2015, includes the opening of at least 22 single-brand shops around the World.

UM Collezioni

Design Agency:
AS Design Service Limited

Design Team:
Four Lau (Creative Director), Sam Sum (Art Director),
Twiggy Yau (Interior Designer) & Francis Wong
(Assistant Interior Designer)

Client:
World First Holdings

Location:
Winman Mall, Macau, China

Area:
126 m²

Photography:
AS Design Service Limited

"SUITCASE" Floating Along the Fashion Stream.

The well-defined positioning of the Female Multi-Brand Concept (MT) will pull together a large customer base for all brands and enhance the image of each brand, creating synergy for all.

Designers Four Lau and Sam Sum are imagining fashion represent sea, using waves as an inspiration element to create a unique shape of island display unit. With the extensive experience in searching and collecting renowned luxury brands among the sea of fashion, UM Collezioni always pick the most fashionable pieces, pack and bring them back in their SUITCASE. This makes UM Collezioni standing in the leading position as always.

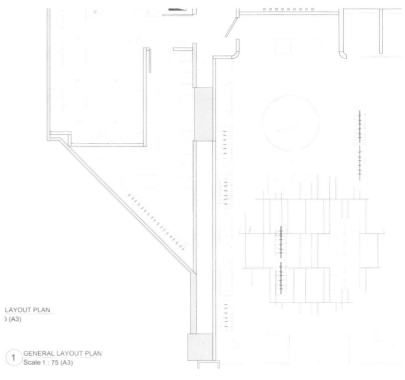

LAYOUT PLAN
5 (A3)

1 GENERAL LAYOUT PLAN
Scale 1 : 75 (A3)

U-LONGE

Design Agency:
Design BONO

Client:
Hyun-Dai Department Store

Location:
Hyun-Dai Department Store,
Mokdong, Seoul, Korea

Area:
165 m²

Photography:
Pyo-Junlee

U-LOUNGE is at the third basement-level of U-PLEX of Hyun-Dai department store. It's a membership lounge for V.I.P of U-PLEX. (The shopping mall for young people of Hyun-Dai department store) They are young and inventive trendsetter from 18 to 35-year-old, called Digital nomad.

To satisfy them, U-LOUNGE provides several exhibitions of limited edition (NIKE, Jeremy Scott, Swarovski, etc.), an abundant variety of colorful attractions and extraordinary experiences.

The member can get drink from the vending machine and use multimedia over the special coin of U- LOUNGE. In addition, this space has a special exhibition monthly on the U-WALL.

"The foggy lake for sunset, branches of large trees swayed in the wind." U-LOUNGE holds the scenery like this.

The facade covered of gradient sheet looks as if landscape photography about the tree on the lake in the fog. The main artwork (tree) expresses the balance of digital and nature metaphorically and unprocessed wood, gradation window film, exposed mass concrete and small lights are shown as the twinkling stars on the foggy lake.

The interior space has various functions, such as cafe, DP (gallery), IT zone.

Designer's intent is communication of nature and digital through a reinterpretation of the material and physical properties.

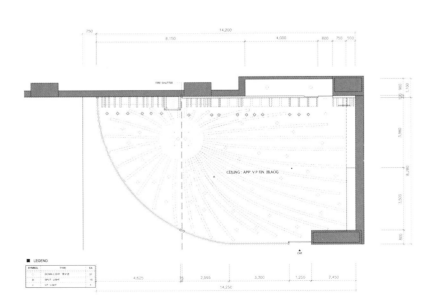

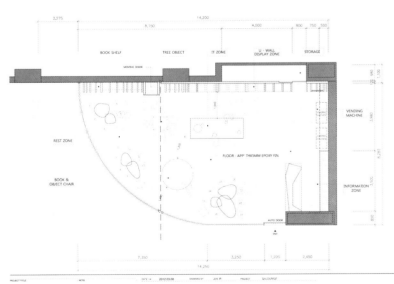

Nature Factory

Design Agency:
Suppose Design Office

Designer:
Makoto Tanijiri

Client:
DIESEL JAPAN Co., Ltd.

Location:
Tokyo, Japan

Photography:
Toshiyuki Yano / Nacasa & Partners Inc.

Denim, a formerly recognized work clothes, has shown different expressions as fashion items to the people. Equally, a group of plumbing, which is usually unnoticed, shows completely different expressions under the name of "Nature Factory". The complex plumbing, trailing by the wall in all directions will cover all over the space. It is like a tree grown over a long time. An atmosphere like a natural arbour is created in the space covered with artificial plumbing.

New attractive scenery is presented with plumbing and fashion-items to show such primarily functional things actually are more diverse and also have higher value.

Raleigh

Design Agency:
OMA

Design Team:
Patrick Hobgood, Darien Williams
with Nicolas Demers-Stoddart

Client:
Raleigh Denim

Location:
Nolita, New York, NY, USA

Photography:
Philippe Ruault

Since 2008, Raleigh Denim has been on a mission to make the ideal pair of jeans using local materials and traditional construction methods from its hometown, Raleigh, North Carolina. In comparison to the increasingly fast paced, global nature of luxury fashion and garment manufacturing, Raleigh Denim is focused on local collaboration, immediacy between buyer and maker, and quality over scale and speed of production.

Using vintage sewing equipment, selvage denim, and knowledge inherited from the region's textile history, Raleigh Denim makes 150 pairs of jeans a week (compared to the 50,000/week produced by a typical factory). Raleigh measures craft in terms of the enduring quality of its product, the community engendered by making, and the personal exchange between seller and user. Its current store in North Carolina, "The Curatory", is an annex to its actual workshop, allowing customers views into the production scene. More akin to a tailor's shop than a conventional retail environment, the store experience revolves around custom fitting, during which jeans are hemmed immediately in front of the buyer to his/her liking.

Raleigh Denim opened its first New York store on Elizabeth Street in Nolita. The space combines two retail typologies – specialty denim and "curatorial" boutique – featuring Raleigh's classic cuts and seasonal collections alongside the fruits of their collaborations in jewelry, shoes, accessories, and curiosities. Inspired by Raleigh's enthusiasm for collaboration and making, OMA designed a flexible display framework that allows Raleigh's team to reinvent the store's atmosphere on a seasonal, monthly and daily basis. The framework consists of a steel wireframe structure onto which apparel, furniture, curiosities, and temporary installations can be easily mounted. OMA has designed a set of clip-on, hanging, and found-object display components that can be attached to the wireframe and rearranged depending on the store's use as a retail, event, or showroom space. The grid is both a functional display solution and an armature for spontaneous creation.

The wireframe divides the existing 87m^2 footprint into three zones – a window display foyer, a denim and collections area, and a "parlor room" for fitting and hemming. A material palette of powder coated steel, custom metal hooks and hanger bars, mirror, pickled plywood, wood-block printed wallpaper, and found objects facilitate both utility and play.

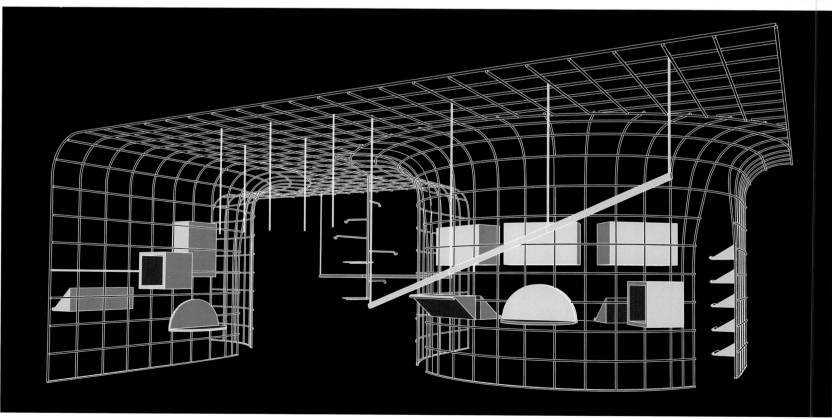

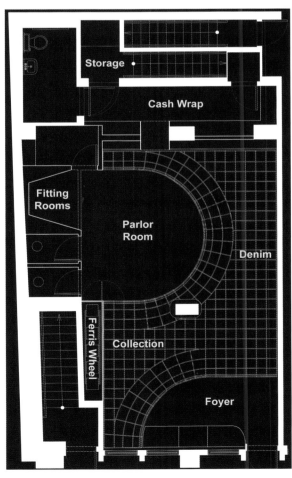

Storage

Cash Wrap

Fitting
Rooms

Parlor
Room

Denim

Ferris Wheel

Collection

Foyer

Heirloom Luxury Bag Store

Design Agency:
Dariel Studio

Design Team:
Thomas Dariel

Location:
Xintiandi, Shanghai, China

Area:
50 m²

Photography:
Derryck Menere

In December 2011, Heirloom revealed its new concept store to the avid shoppers and fashion-goers of Shanghai, in the heart of Xintiandi Style mall. The design concept is a first of its kind, which highlights the brand's full collection of leather accessories. Building up on Heirloom, the new space is an extension of its existing concept store, with a whimsical twist.

The main challenge faced by the designer was the size of the store: 60 squaremeters. The concept behind the store is to re-create as pace where the customers feel like they have entered a contemporary yet elegant boudoir. Upon crossing the classic metallic gold entrance, the visitor is instantly drawn into a fantasy world.

Immediately, the black and white striped marble flooring reflecting into the reception desk creates a new perspective to enlarge the space.

The Art deco grey colored walls showcases a selection of exclusive handbags that are recessed in white lacquered cloud shaped frames. These frames, an HEIRLOOM signature design are upholstered in a creamy delicate fabric for a luxurious feel.

The interior decorations are exclusively designed for the space.

The highlight of the new store is the exclusive boudoir nested in the corner of the space, quietly tucked behind a stainless steel gold dome. This exclusive space is the quintessence of refinement with its embedded Heirloom signature mesh studs scattered across the wall. The luxurious touch invites the clients to discover the products as jewels and to re-create an intimate and pampering shopping experience.

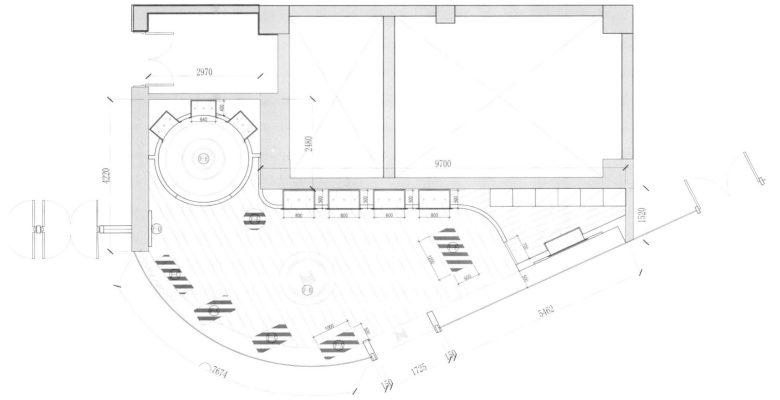

AMERICAN RAG CIE Namba

Design Agency:
MOMENT Inc.

Design Team:
Hisaaki Hirawata
Tomohiro Watabe

Client:
AMERICAN RAG CIE
JAPAN INC.

Location:
Namba Osaka, Japan

Area:
279 m²

Photography:
Nacasa & Partners

The corridor is the main design concept to make use of the space with a long depth. AMERICAN RAG CIE is a selective product boutique suggesting a free lifestyle.

The customers find a variety of fine goods along the corridor. The memorable arches in a silhouette softly divide the shop into three space: men, ladies clothes, and household goods.

The interesting different shapes of arches attract the customer's attention because of the colors, shapes, materials and the products they will find on the display. The vintage decoration wall finishing, expresses the brand's philosophy and unlimit free selective style. The pure simplicity of the minimum elements at the entrance is significant. At first glance, it looks graphically flat from the facade, but once you walk in you will see something different.

Max Mara Flagship Store in Chengdu

Design Agency:
Duccio Grassi Architects

Location:
Chengdu, China

Duccio Grassi Architects designs the Max Mara flagship store in Chengdu, on Renmin South road. The 720 m^2 articulated on two floors makes it one of the biggest Max Mara shops worldwide. The flagship designed by Duccio Grassi Architects has a narration through diversified experiences, where the floors become new attractive spaces for the visitor.

An important element is the innovative outside membrane made of burnished and brushed copper which characterizes the main entrance: the façade is composed of three levels of burnished copper which hide the lighting system that allows the store to change its face at night.

The material and the light define the design of the façade and set it apart from its surrounding. The secondary entrance located inside the mall, in opposition to the main one, presents a more essential pattern: large shop windows alternate with simple surfaces made of burnished copper that frames the steel structure, designed as a semi-transparent and lighting wall, which partially screens the inside space through thin septums made of opaline lighted glass, wood and platinum-glass.

The contraposition, which is an essential theme of the project and defines the rhythm of the extended boutique, continues in the selection of materials, which are combined in an unexpected way, reflecting the unusual combination of fabrics in the new Max Mara collections.

Wood is the absolute protagonist: ash-treated oak, rattan and recycled wood communicate the continuous blending of cultures and materials. The contrast is well-balanced with the hard materials selected: platinum-treated glass, burnished copper plates, shiny stainless steel.

On the ground floor the theory of small size products and accessories takes place near a versatile space for multi-functional installations that can change from one season to another. A protagonist of this floor is the staircase, with a vertical wall made by horizontal panels padded with soft pink fabric, which leads to the upper level of the store.

Galeries Lafayette Jakarta

Design Agency:
Plajer & Franz Studio

Location:
Jakarta, Indonesia

Client:
Galeries Lafayette &
Mitra Adiperkasa

Area:
12,000 m²

Photography:
diephotodesigner.de

Galeries Lafayette, the legendary department store and symbol of French art de vivre, had finally arrived on the Indonesian retail horizon. Located in Pacific Place, one of the most up – and coming areas in the heart of Jakarta, the department store is an inspiring fashion temple. More than 300 international brands spread on four floors reflecting the classic and modern french chic.

The exterior design and store concept have been developed by Berlin based architecture & design studio plajer & franz studio with the idea of bringing a touch of parisian feeling to Indonesia.

One day in Paris… Is the leitmotif of the design concept for Galeries Lafayette in Jakarta. The unrivaled flair of the French capital is visible and sensible in every corner of the 12,000 m² large space. Typical parisian elements such as building facades and the Eiffel tower are interpreted in an accessible but warm way. Plajer & franz created a modern fashion temple that reflects the "mix and match" philosophy of the Galeries Lafayette brand.

A modern interpretation of the Eiffel tower presented as a huge vertical element next to the staircase, makes an impressive and powerful statement right at the entrance. Unlike in Europe, the landmark of the French capital enjoys great prestige in Asia and brings about a highly estimated value. The "Coupole" – an emblem of the great architecture of the parisian department store – is also respectively celebrated within the store design for Galeries Lafayette in Jakarta. It reapears as an over-sized ceiling and lighting element on all four floors marking the crossing points between various segments and highlighting the DNA of the brand.

A further creative link to Paris is reflected by the so called "indoor-facades", which double-sided indoor walls are made of sandblasted glass and mirror glass. These reflect the French flair by displaying alienated sketchy motives of parisian building facades and streetlamps. At the same time, they have the role of separating various brands from each other and in this way creating a kind of display window for each one of them.

The materials used in-store are selected adequately in respect of the varied segments – from casual to luxury. For the various elements reflecting the brand's DNA and creating a brand statement the plajer & franz design team used a lot of bronze and printed glass. However, red as one of the CI colors of Galeries Lafayette draws a continous line throughout the entire department store making its strongest statement in the Eiffel tower. Natural stone and oiled wood have been used for the floors creating a natural look while beeing robust accordingly to the demands of a department store.

Generally the store architecture creates a scene for the product and the brand "Galeries Lafayette". Its DNA, identity, mission and vision are put to the fore while setting a framework for the multiple and diverse sub-brands.

The Galeries Lafayette department store in Jakarta has emerged in a partnership between Galeries Lafayette and Indonesia's leading fashion retail company Mitra Adiperkasa (MAP).

Ellassay Retail Design Store

Design Agency:
Studio 63 Architecture + Design

Location:
Guangzhou, China

Area:
130 m²

Interior design:
Studio 63 Asia Ltd.

Photography:
Studio 63 Asia Ltd.

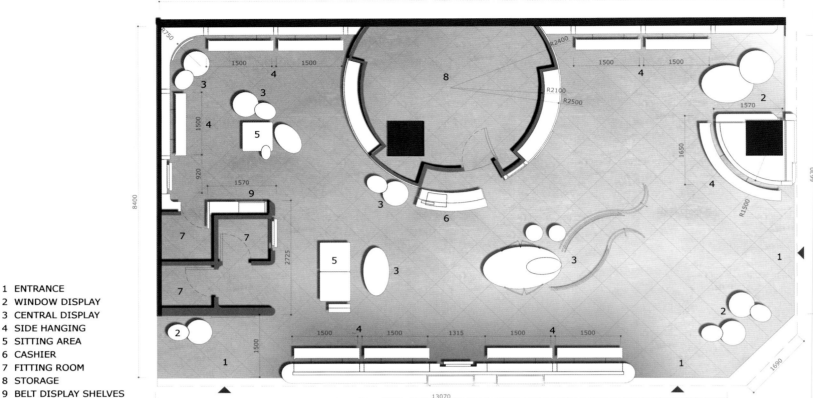

1 ENTRANCE
2 WINDOW DISPLAY
3 CENTRAL DISPLAY
4 SIDE HANGING
5 SITTING AREA
6 CASHIER
7 FITTING ROOM
8 STORAGE
9 BELT DISPLAY SHELVES

Studio 63 Architecture + Design was called upon by Ellassay to design the new store image in China. From Shanghai to Beijing, the footprint of Studio 63 is always presented in the graceful world of Ellassay.

The core values of Ellassay are condensed in the symbol of the rose: elegant, gentle, classic and luxurious. After thorough research and discussion, Studio 63 and Ellassay had agreed to take the symbol of rose as the starting point for the new concept design.

The new concept of the Ellassay store revolves around the rose core. It is made with different layers of wooden lines of different thickness and bronze in rose gold as the background. The main idea is to create a stylized rose with layers of petals where clothes can be displayed upon.

This rose core is modular and can function as mannequin display and cashier. The space inside the core acts as storage. It is an interesting fusion between beauty and function.

The features of the rose core are repeated around the whole store. The horizontal wooden lines are extended around the store which echoes the intertwining petals.

The central elements are inspired by the freshly growing rose buds where part of the petals is beginning to bloom. Stylized and modern, these central elements seem smiling. They serve as mannequin stands or display table where Ellassay products are preciously displayed.

Soft lighting is used to create a warm and welcoming environment. The lighting effect is reinforced with the use of shiny material such as bronze and glossy white paint. As a result, a bright and elegant store is created to represent the luxurious world of Ellassay.

ESTNATION
Nagoya

Design Agency:
MOMENT Inc.

Design Team:
Hisaaki Hirawata,
Tomohiro Watabe

Client:
ESTNATION INC.

Location:
Nagoya, Japan

Area:
387 m²

Photography:
Nacasa & Partners

ESTNATION mainly sells the clothes for both sexes. What we attempted is leaving the general idea, which means that the shop is normally divided into some zones by the walls. Here, such various furniture would draw the attention of customers, as sculptures, metallic blue pillar, glued laminated timber and sharp facade vaguely partitoned in the whole space. The simplicity in colors and space would create a strong appeal leading all the passersby to the shop, then they will be easy to walk inside without any wall partitions. What is important for the retail design is that letting them walk a lot and see the commodities to stimulate customers' psychology. It would be dull to divide clearly the space into ladies' and men's area by the walls, but the presence like a sculpture guides them and show where they are inside the shop with interest and curiosity.

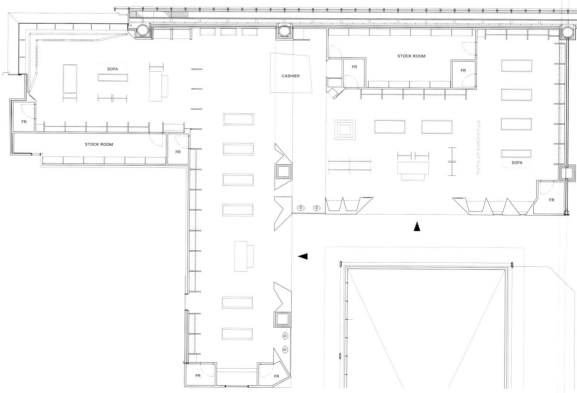

ALV
Showroom

Designer:
Fabio Novembre

Design Team:
Lorenzo De Nicola,
Giuseppe Modeo,
Patrizio Mozzicafreddo

Client:
Alviero Martini

Location:
P.zza San Babila n°5,
Milan, Italy

Photography:
Pasquale Formisano

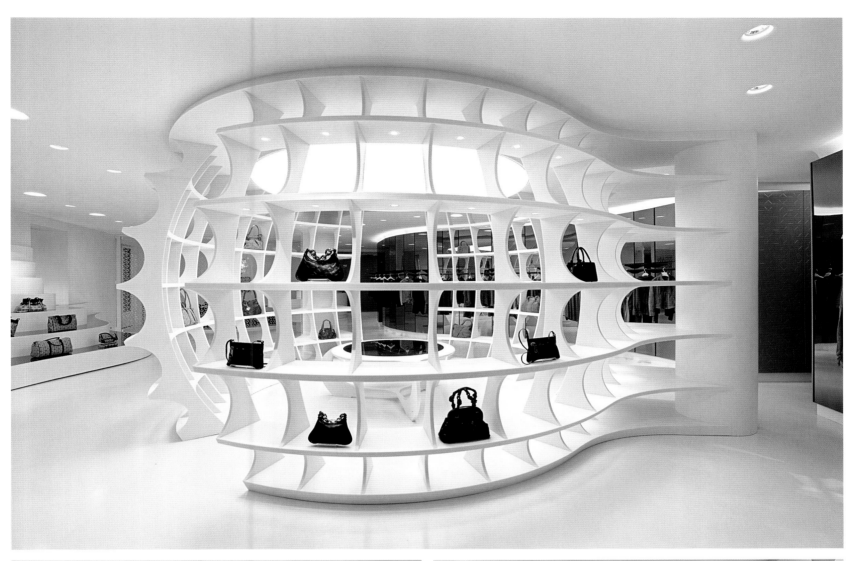

Go, far away, travelling. This is the philosophy that characterizes the new professional era of Alviero Martini.

Travel has always been the core of his world, while during the years, it has become a journey without any point of reference, an exploration with the aim to go far away through the closest.

The ethernal and the unchangeable planets movement has been represented through a huge white atom, the marvellous Time and Space machine able to transform the concentration of material in pure expansion.

The space becomes a fluid interchange of centrifugal surfaces and centripetal reflexs, a balance between freedom and dependence based on the sharing of a common project... as life is.

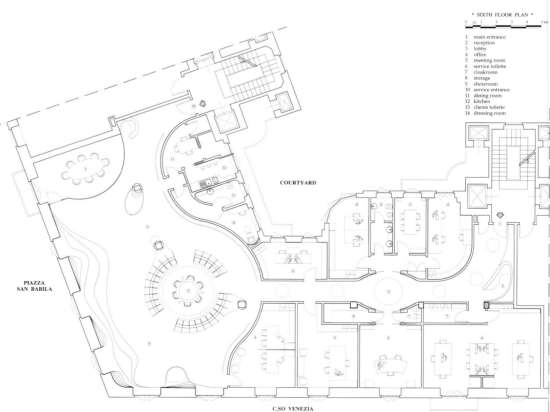

* SIXTH FLOOR PLAN *

0 1/2 1   2   3   4   5 m.

1   main entrance
2   reception
3   lobby
4   office
5   meeting room
6   service toilette
7   cloakroom
8   storage
9   showroom
10 service entrance
11 dining room
12 kitchen
13 clients toilette
14 dressing room

COURTYARD

PIAZZA
SAN BABILA

C.SO VENEZIA

New Extension
Building for "Foeger
Woman Pure"

Design Agency:
Pedrocchi Architects, Basel

Designer:
Reto Pedrocchi

Client:
Ms Midi Foeger, Telfs, Austria

Location:
Telfs, Austria

"Foeger Woman Pure" has been one of Austria's leading fashion boutiques for the past 20 years. Pedrocchi Architects from Basel, Switzerland, extended the existing fashion boutique with a cupola-like building which houses the collections of new and promising designers.

The new extension building is planned as a huge showcase with continuous transition of external and internal floor covering.

The formal requirement for the extension building was to continue the two-dimensional gabled roofing of the Tyrol houses in the neighborhood. The roof of the extension, however, has almost nothing to do with conventional gables. It's rather a geometric system of entangled triangles that span an area of 140 square meters, which features an arch without any supports.

The internal room is dominated by two massive concrete beams which span the room diagonally and form an x-shaped cross at half the ceiling's height. Although these beams are required for structural reasons, they are both used functionally and formally. Oversized stainless steel hooks are applied to these beams. They will be used for hanging up and display exclusive fashion garments. Customers can stroll through the fashion boutique and feel like being on a stage.

Pure materials dominate both the interior and exterior of the new building. The roof is clad in black perforated plate; the internal space is constructed from exposed concrete, glass, chromium steel and a light-colored natural stone floor. Slightly curved walls covered with shimmering glass mosaics separate the changing rooms from each other. Every single cubicle is flooded by natural light from an overhead skylight.

Tangy Store

Design Agency:
OOBIQ Architects

Location:
Tianjin, China

Tangy Collection Store in Tianjin has been opened for a few days. Play of design is a new interpretation of artistic creativity.

Designers followed artistic instinct and presented a combination of the past and the present, making it a place where the urgent rhythms of modern city life mix with the tranquility of primitive life. The wooden display stages, simple, modern and unique, show up to advantage with the ribbon hanging. In addition, the netlike partition created with the special type of silk cultivated in Tangy's factory, together with the endless ribbon made again of silk, look unique. They are from the present, but they stir the imagination, stimulate the senses and remind of something from the primitive society.

A poetic approach to interior design allows your imagination to spin off on tangent. Elegance is in the Italian blood. The designers blended this feature with the spirit of oriental culture-minimalism, low profile and hand-made human touch. The simple structure and lines, which are inspired by Zen, respond to the sensitivity of people in China. It cultivated a modest, practical but also intimate style.

仓库
STORE RM.

試衣間
FITTING RM.

試衣間
FITTING RM.

收銀台
CASHIER

Shang Xia
Beijing Store

Design Agency:
Kengo Kuma &
Associates

Client:
Shang Xia

Location:
Beijing, China

Area:
152 m²

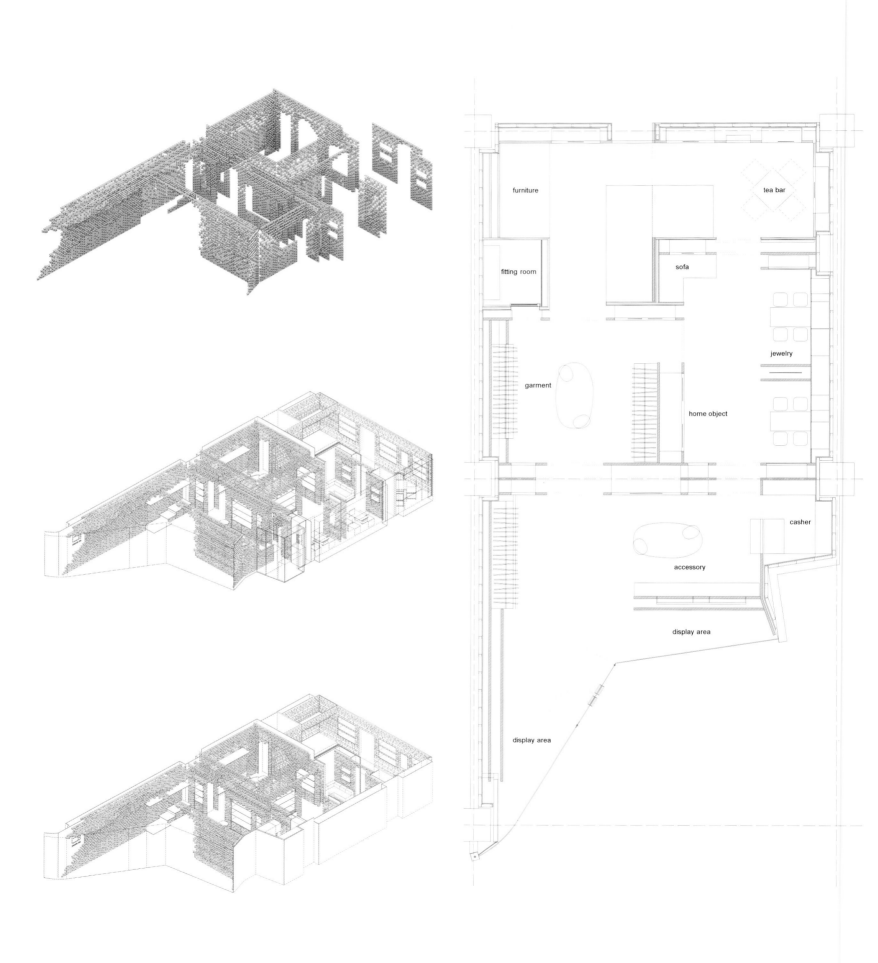

furniture

tea bar

fitting room

sofa

jewelry

garment

home object

casher

accessory

display area

display area

In this shop, extruded aluminum is used as the main material to form the space. The aluminum consists of three H-shaped types (H: 60mm, H: 90mm, H: 135mm) and two L-shaped types (L: 100mm, L: 200mm). At the upper and the bottom part of the space where the load is concentrated, the shorter type (H: 60mm) of aluminum is densely applied. To the contrary, the higher type (H: 135mm) of parts is used largely in the middle, as the load is less, so the screen could be light. Thus, feature of this design is virtually the result of the structural demand, but the mechanics naturally generated a gradually-changing transparency from the material. The layer of the aluminum screens makes you feel being placed in a mysterious cloud.

Designice Hangzhou Flagship Store

Design Agency:
Guan Interior Design Co.,Ltd.

Designer:
Zhang Jian

Client:
Designice

Location:
Hangzhou, China

Area:
1,120 m²

Photography:
Wang Fei

1F.1

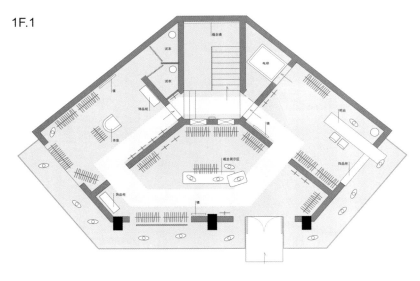

2F.1

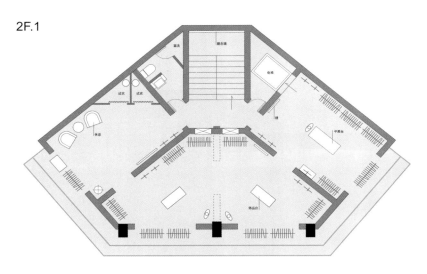

3F.3

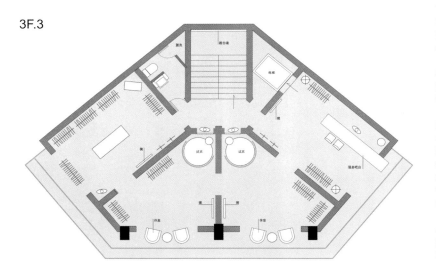

The new brand profile of Designice 2010 employs the design concept of grid for the building's exterior walls. The grid element has its application from this 7-storey building's top down to the bottom, making it as a unified whole in its exterior and at the same time veiled the windows, air-conditioners etc to have a better look.

Storey 1-3 is the zone for retail, 4-6 for office and training function, 7 is the DS clubhouse. When lighting up, the interior space, hazy but cozy and separated, brings a fresh experience for the customers as well as the proprietors.

Due to the complicated internal structure of this building, it is challenged for the designer to lay out the dispersive available space reasonably. Considering from the building itself, the designer chooses light color as the color base. White applied for the walls and ceiling, brightens the whole space and creates a spatial and comfortable shopping environment. White tiles for the main parts and accompanied by the solid wood flooring which is paved along with the design generatrix, the floor can well lead the customers to walk the right path.

As a solution to the narrow internal space of the building, glass is applied in some important space to enlarge the air permeability and improve the comfort level, helping to retain the customers for a long stay.

Bynamesakke
Fashion Store

Design Agency:
Inside/Outside

Designer:
Agi Kuczyńska

Area:
32 m²

Location:
Warsaw, Poland

Photography:
Jakub Certowicz

bynamesakke

This is the first flagship store of the Canadian-Polish clothing brand Bynamesakke. The project was developed in close cooperation with the brand designer, Iza Keyes – Krysakowska and the architect Agi Kuczyńska. The project aimed to meet the brand concept, which focuses on natural materials, craftsmanship, simplicity and material quality. Shop was located in "Złote Tarasy" shopping center in Warsaw, so the project had to be adapted to the complex requirements of the center. Small area and a long, sloped shape of the plan required the opening of the store towards the passage and optical extension of the store. The idea was to create a bright friendly interior made from natural materials that would be complementary to the brand identity. The client wanted to have more of a cosy feeling then a flashy shopping mall space. The space was symbolically divided into two parts: shorter side painted white, where there the racks display classic cut shirts, basic T-shirts and tank tops and the longer side-wood cladded, where the seasonal collections are shown. Oiled oak has been used on walls, floors and furniture in order to give the interior a warmer feel. White ceiling is made from a special non-flammable paper structure that evokes the repetition of the fabric waving. Lights located above the paper ceiling create a shadow play evoking the light and shadow play of the sun shining through the branches of the trees. This ambient lighting is supplemented by spot-lights directed at the hangers. In order to emphasize the lightness and airiness of clothing, stainless steel hangers were hung from the ceiling not attached to the floor.

C&A Brazil: Iguatemi Flagship

Design Agency:
Chute Gerdeman

Location:
Sao Paulo, Brazil

C&A Brazil's new flagship store in Sao Paulo's Shopping Center Iguatemi takes the fast-fashion retailer's brand to new heights of glamour and luxury.

The store creates an entirely new C&A brand experience for the Iguatemi shopper — elevating the in-store environment through luxurious material finishes, arresting visual presentation, iconic architecture, compelling brand messaging and unique customer services, all geared to the targeted Iguatemi female shopper.

"C&A Brazil entrusted us with their brand in this historic retail destination in Sao Paulo, and we couldn't be more proud of the result," said Jay Highland, Chute Gerdeman's Vice President, Brand & Marketing. "From the beginning, this project demanded the expertise of each of our disciplines, with brand strategy guiding all of our decisions, through design and implementation."

Drawing inspiration from the soul and spirit of the Brazilian Woman (Espirito Brasileiro), Chute Gerdeman created a store drenched in light, energy and happiness. High-fashion and affordable luxury combine to create an illuminating entry to Brazil's most prestigious retail shopping mall.

At the heart of the store is a wide, circular staircase, which draws the shopper up through the space with a one-of-a-kind digital fashion-show wall. The sensational, moving LED display is a three-storey architectural marvel, wrapping the staircase in constantly changing light, movement and energy.

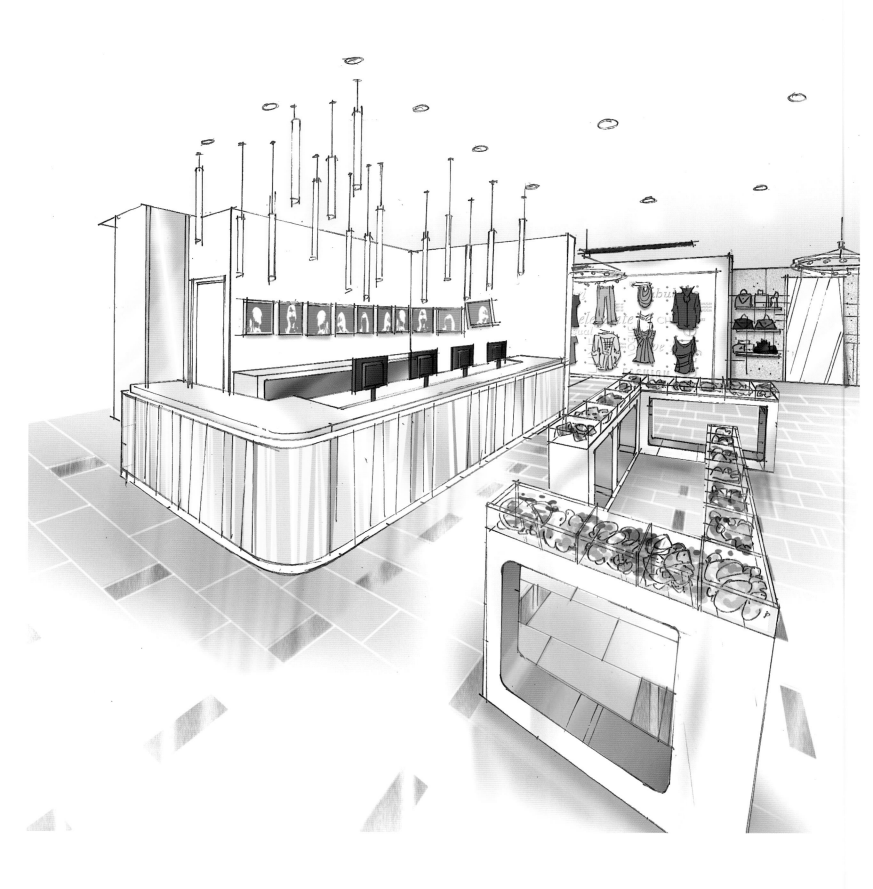

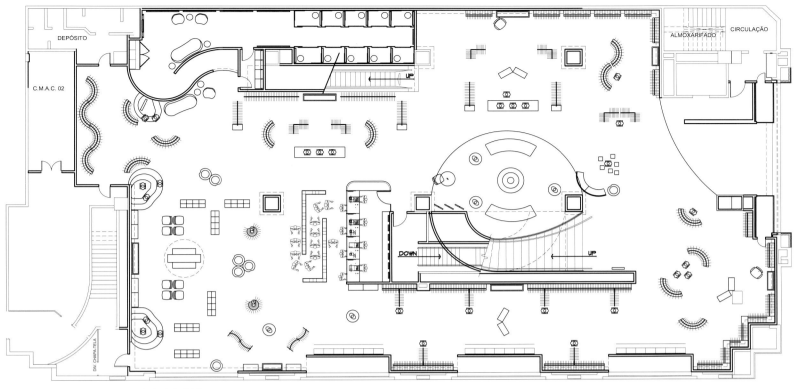

GROUND FLOOR PLAN
DATE: 02.05.10

MIDDLE FLOOR PLAN
DATE: 02.05.10

C.M.A.C. 03

SANIT. MASC.

SANIT. FEM.

OPEN TO BELOW

OPEN TO BELOW

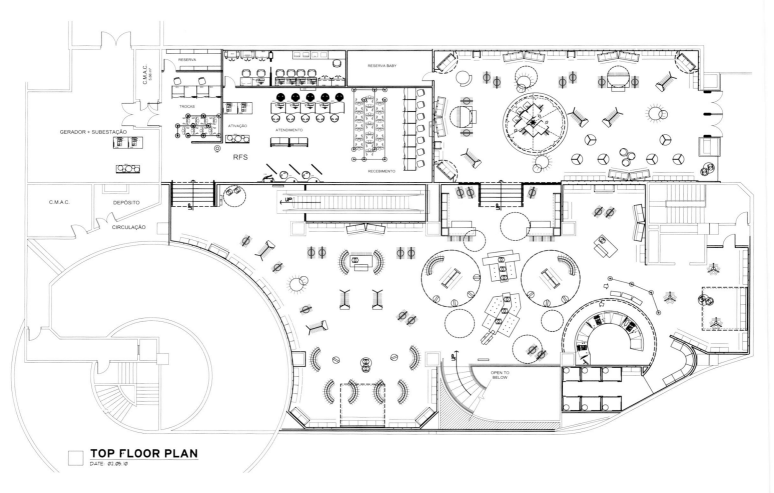

TOP FLOOR PLAN
DATE: 02.05.10

SPIRAL in Yinchuan

Design Agency:
SAKO Architects

Design Team:
Keiichiro Sako,
Satomi Nagaoka

Client:
Yinchuan Yagaoshangdu
Trade Co.,Ltd.

Location:
Yinchuan, China

Area:
335 m²

Photography:
Ruijing Photo

This small mall is composed of 5 boutiques, 2 shoes shops, and a VIP room. In the center, there is a stainless steel mirror finished spiral staircase, and the planning creates an ease of getting around. The mass of the metal staircase is made by hand, and its surface creates sullen and glossy reflection. In the dim lit room, the staircase shows the unique presence.

Shine Fashion Store

Design Agency:
NC Design &
Architecture Ltd.

Designer:
Nelson Chow

Client:
Shine Trading (HK) Ltd.

Location:
Causeway Bay,
Hong Kong, China

Photography:
Dennis Lo Designs

Shine is one of Hong Kong's most renowned high-end multi-brand fashion stores, known for bringing pioneering foreign brands to the trend conscious locals. For the second shop located in the high traffic youth-oriented shopping district of Causeway Bay, the owner specifically requested for NC Design & Architecture Ltd. (NCDA) to produce a design that would reinforce the company's identity as an avant-garde and experimental fashion store.

Inspired by the name of the store, a 7m tall asymmetrical glowing star-like structure forms the primary street identity along Leighton Road, attracting both pedestrians and motorists. The pristine white shell embodies a black interior wall that further unfolds to create three main rooms: The entrance gallery, the upper level sales area & finally the dressing room. Equipped with 3 display platforms and suspended mannequins, the entrance gallery acts as an extension of the window display and forms a stage for the evolving seasonal Merchandise displays. The crystalline black wall unfolds to form a suspended stair leading to the upper level sales area, and a row of geometrically arranged fluorescent lights is placed above the stair to emit a cool futuristic sci-fi glow which goes in line with the progressive spirit of the clothing.

The upper level sales area showcases the men's and women's ready-to-wear collections in the black crystalline niches on both sides. Special attention is given to the display of the latest pieces, which are suspended on two central uplit racks. Pieces from various designers are presented against a monochromatic background. A continuous metal edge above each niche allows for the flexible placement of magnetic brand tags in order to showcase the evolving selection of designers.

Finally, the dressing room conceals the leather padded fitting rooms and cashier entrances behind a continuously folded kaleidoscopic mirror partition, forming the most intimate and private area within the overall shop. Inspired by music videos and computer generated effects, the dressing room enclosure creates a "hyper-real state", where the customer can see multiple reflections of themselves at different angles in the mirror. The back lit stretched ceiling creates a false sense of depth to the 2m headroom yet provides abundant light to the person trying on the clothes.

The design of the Shine flagship store in the Leighton Center showcases how the idea of a "shining star" could be translated architecturally into a fashion retail space, creating a visually striking yet highly functional contemporary store.

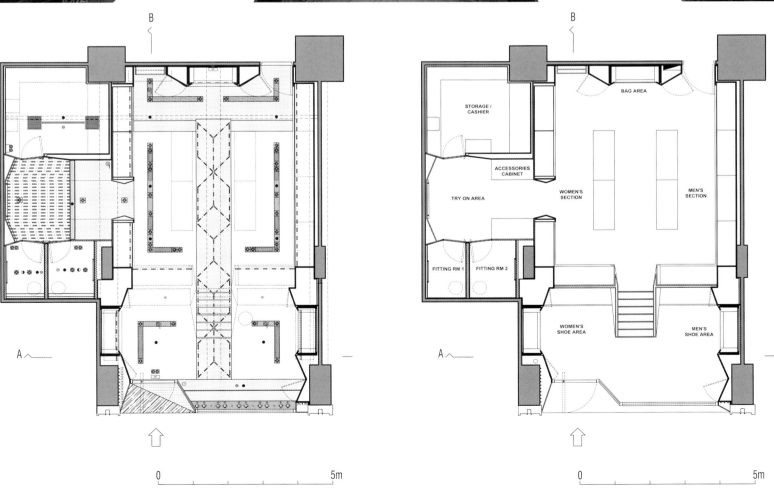

B

B

STORAGE / CASHIER

BAG AREA

ACCESSORIES CABINET

TRY ON AREA

WOMEN'S SECTION

MEN'S SECTION

FITTING RM 1

FITTING RM 2

A

A

WOMEN'S SHOE AREA

MEN'S SHOE AREA

0   5m

0   5m

Shine Fashion Store "A Store of Infinite Curves"

Design Agency:
NC Design &
Architecture Ltd.

Designer:
Nelson Chow

Client:
Shine Trading (HK) Ltd.

Location:
Guangzhou, China

Photography:
Dennis Lo Designs

As one of Asia's most renowned high-end multi-brand fashion stores, Shine is famed for bringing pioneering foreign brands to the trend conscious locals. For its third store located in La Perle, Guangzhou, NCDA produced a design that reinforces the company's trend setting identity. The Shine La Perle store reflects the drapery technique. This curving scheme omits any abrupt corners within the store, making the 167 m^2 store to look wider whilst enhancing the overall flow of circulation and creating a sense of fluidity.

The fluidity of the drapery technique is apparent from the glass exterior, seamlessly leading into and throughout the store, morphing from glass to metal to wood. The entire store, linked by one continuous black veneer wood drape, unfolds to showcase various designers' collections.

At the entry, customers are greeted by an organic shaped central platform that acts as focal point of the store and as an extension of the window display, creating a gallery like space where the display changes seasonally. A large round stretched ceiling drops down, shining on the display platform below while a series of full height bronze screens lie behind the central platform, further framing this as the focal point of the store. Taking on a sculptural form, the screens also act as concealed doors to the intimate leather wrapped changing rooms beyond.

Follow the curves on either side of the central platform will lead customers into men's collection on the left and women's collection on the right, the curving walls unfold from one brand into another, creating a fluid circulation experience while still physically separating the various collections.

Taking a closer look at each display niche formed between two curvilinear pillars, one will discover that these curved wood pillars also conceal the lighting systems for the merchandise. To allow the continuous flow of light, the shelf within each niche is cantilevered from the back and does not touch the sidewalls, thus allowing light to shine through uninterrupted. Each niche uses a versatile rack where merchandise can be displayed with a front or side profile. A shelf placed above each rack allows each brand to showcase its clothing and accessories simultaneously, utilizing the full height of the 3m space and enhancing the overall presentation of each collection. Adhering to the operational needs of the store whilst staying in line with its monochromatic theme, NCDA has designed custom grey hangers and custom magnetic name tags that can be placed anywhere along the metal edge above the rack, thus suiting the varying size of each designers collection while enhancing the brand identity of Shine.

The curvilinear pattern is echoed in the stone / carpet flooring and mirrored in the ceiling light trough that contains the architectural lights and mechanical ceiling grilles. Thus forming a seamless curvilinear design that blends the wall, flooring and ceiling into one unified whole. The fluidity seen in the drapery of a black dress forms the inspiration for Shine La Perle. Partnered with high functionality in a contemporary style and staying true to its design philosophy of light and fashion deconstruction, Shine Guangzhou is a store of infinite curves creating a truly unique experience for customers.

L'ECLAIREUR
Fashion Boutique

Design Agency:
SAQ Architects

Designer:
Roel Dehoorne

Client:
L'ECLAIREUR

Location:
Paris, France

Area:
450 m²

Assistant:
Isabelle Speybroeck

Inspired by a friend's dressing room, the client wished visitors to set foot in the new space as if they were entering an intimate room in the midst of the hustle of the city.

Viewed from the streetporch, nothing suggests this new public space other than a long corridor whose slight downward slope and slanted walls infuse the viewing angle with a dramatic effect.

Inside, the gaze is immediately drawn by an installation built in situ by artist Arne Quinze, a weave of cleats and wooden slats balanced out in a state of utmost tension.

One then discovers a series of spaces outlined and shapen by a meandering wall. As if carved out, the irregular contour of the wall daubed in a glossy champagne colour displays in controlled disarray the items in a series of alcoves, niches, or secret room compartments.

The walls are materialized by an assembly of recycled elements composed of wooden planks, aluminium printplates and cardboard sheets. At first glance, they appear to be fragile, almost friable. A mere touch by hand shows the contrary to be true. With a technique that is normally used in automobile industry, the ensemble has been given a thick polyurethane coat, lending the walls all the solidity they need. A multi-layered finish of lacquer and varnish give the ensemble a velvety gloss. Surprisingly, a sense of luxuriousness exudes from pieces that are best described as "trash".

Arriving in the spacious choir underneath a glass canopy at the back of the shop, the physical and the digital world merge to create a baffling spectacle: the array of TV-screens seems to have proliferated and almost drown the space in the light of their projected images. No less than 32 video feeds spread over 100 screens aim to generate a different ambience, adaptable to the diverse organized events.

This dynamic stage setting reaches its climax when the walls open up upon command, revealing exclusive fashion creations till then deprived by the inquisitive gaze of the visitor.

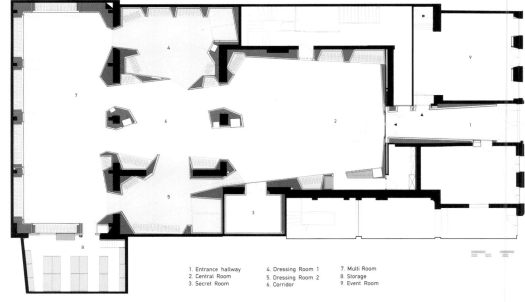

1. Entrance hallway
2. Central Room
3. Secret Room
4. Dressing Room 1
5. Dressing Room 2
6. Corridor
7. Multi Room
8. Storage
9. Event Room

Stylexchange

Design Agency:
Sid Lee Architecture

Client:
Conceptwear

Location:
Montreal, Quebec, Canada

Area:
375 m²

Photography:
Sid Lee

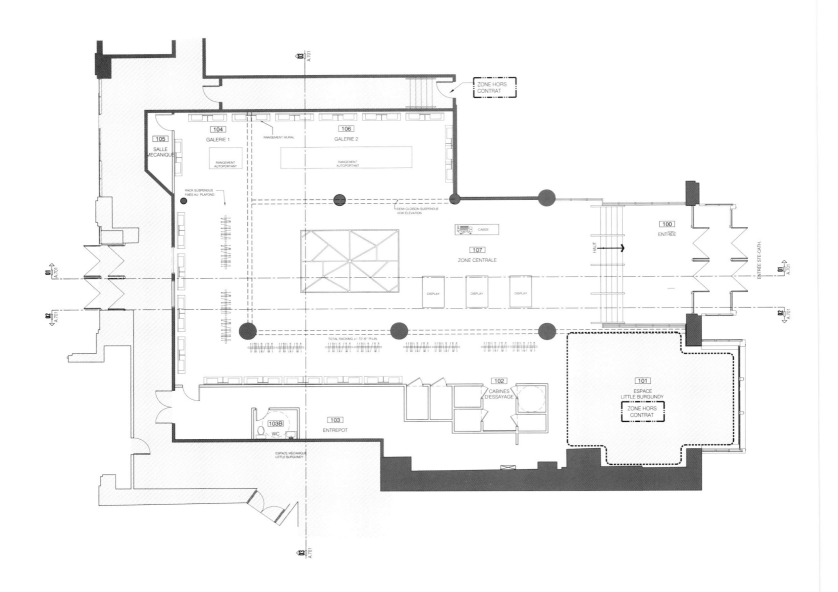

Stylexchange Faubourg is a retail outlet located in one of the entrances of a Ste. Catherine Street historic food mall. Located right next to Concordia University's campus in downtown Montreal – a young and vibrant student neighbourhood – the boutique blends seamlessly into the urban landscape of this diverse, multicultural part of town. The simple and flexible design showcases Stylexhange's contemporary fashions while at the same time recounting the history and traditions of its surrounding neighbourhood.

The black area showcases recently arrived products in a non-organized, market- inspired way. The central workshop, bright and white, is drawn in a way to focus on the work of the in-store stylist who puts together noteworthy combinations of clothes and accessories.

The existing entryway of the store was amended to create a linear commercial space, with low ceilings above an open-plan, two-level space in the centre. The aesthetic of the new atelier is part design studio, part trendy, second-hand boutique. The store was designed as a canvas for the diverse, cosmopolitan crowd of the area: style, music and graphic art are the real inhabitant of this space.

This central white steel structure is built like an unfinished tent. Loosely but strongly illuminated, it dominates the room, serving both as a showcase for new products arrivals and as a musical and entertainment centre when fashion shows are taking place in the store.

The black space was conceived as a creative canvas of blackboards, with a constantly evolving design. The blackboards allow local artists to decorate the boutique in the colours of the surrounding neighbourhood. In this way, art unifies every element of the store, connecting the brands with each other and with the neighbourhood.

The original floors of the space were maintained. They feature a mix of tiles from different eras, while the subdivisions of the space were painted uniformly, thus recalling the history of the space.

My Boon

Design Agency:
Jaklitsch /
Gardner Architects PC

Design Team:
Stephan Jaklitsch, Mark Gardner, Christopher Kitterman,
Mariana Renjifo, Bjarke Ballisager, Margaux Schindler, Liz Kelsey

Client:
Shinsegae Co., Ltd.

Location:
Seoul, South Korea

Area:
622 m²

Photography:
Nacasa & Partners

Designed by Jaklitsch / Gardner Architects of New York, My Boon is a new 622 m² lifestyle boutique on Cheongdam-Dong in Seoul's Gangnam-Gu neighborhood. The basic concept for the shop derives from a simple statement from the client: "My style, my body, my soul." The focus is on the best in trends, fashion, accessories (style), beauty cosmetics and vitamins (body), and art, music, movies and books (soul).

The overall retail experience is structured to be anything but typical – My Boon's ambitions are evident in the fact that the objects, clothing, cosmetics for sale are displayed. This served as a starting point for our design investigation, which sought to create an architecture that would reflect the brand's ethos and the balance between the best of contemporary culture and a natural, healthy lifestyle.

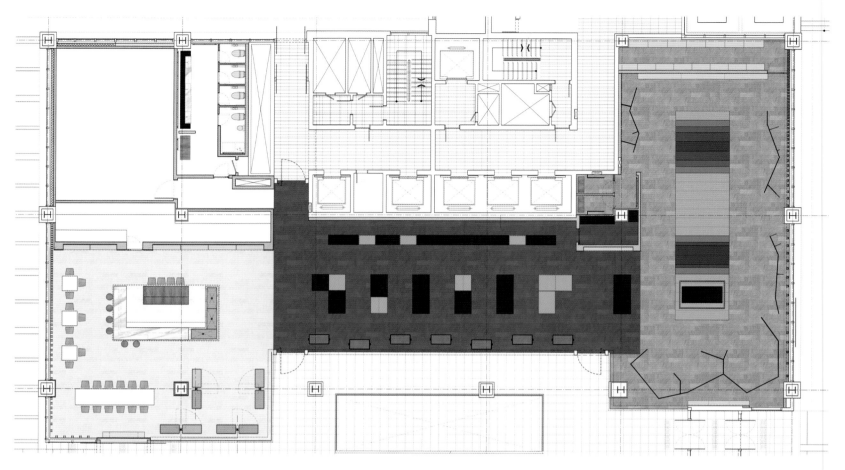

The scheme is divided into three zones: a large retail space for fashion, apparel and footwear; a juice bar and café; and a retail gallery of personal accessories and objects that acts as a connector between the two spaces. The design defines each of these zones through subtle shifts in materiality and color. The project employs unifying elements to tie the spaces together. Wood end-grain flooring is used throughout, with changes in color at each zone: natural finish at the north sales area, black at the retail gallery and white stained finish at the café. To reinforce a visual connection between the three zones, a louvered ceiling of natural wood runs the length of the space. Throughout all three zones, polished, stainless-steel shelving details add a consistent element to hand-troweled concrete walls and reclaimed, rough-sawn wood. Black is used as a framing device in each space to reinforce the spare palette of light woods, troweled concrete, white carrera marble, and polished stainless steel.

Throughout the shop, careful attention was paid to curating merchandise within a mix of social spaces. In the retail area, concrete planks, which feel more industrial, are stacked across the sales floor to create an adjustable display space and impromptu seating areas. The cafe is the main social space of the store, where customers can recharge their bodies with a healthy menu of fresh juices and snacks. It is part apothecary, as well as a café, with a hygienic atmosphere. The café area features a counter-height bar of white marble and gradient white glass above that gives an ethereal lightness to the space. The design is an investigation of the artificial versus the natural – wood surfaces punctuated by rubber tabletops, blackened-steel display boxes and concrete planking. It is not so much materials in opposition, but, rather, all materials having a level of industrial refinement to create a sense of ambiguity and mystery.

MAX & Co
in Hong Kong

Design Agency:
Duccio Grassi Architects

Client:
Max Mara Fashion Group

Location:
IFC – Hong Kong, China

Area:
157 m²

Photography:
Virgile Simon Bertrand

"The product has to be the protagonist of the shop" is the utterance most frequently repeated to me by the clients. In this case the product should be the main object of the design thoughts and "how" it is displayed is the final goal of the design. Instead Duccio Grassi believes an exposing place is primarily composed by the human beings who relate themselves with that very space and see that specific product. An exposing place without visitors or clients has no meaning. He believe that the major thoughts on design have to address to people.

Following this concept DGA designed a space where the product is not visible from the outside but people is attracted by visual stimuluses and suggestions.

The space is prepared with volumes that are pure geometrical shapes, cylindrical with casual textures but also apparently casual shapes closed with geometric severity.

DGA designed big metal volumes resembling a white a light embroiderer and volumes following the concept of shell with the exterior made of burnished brass and the inside painted white.

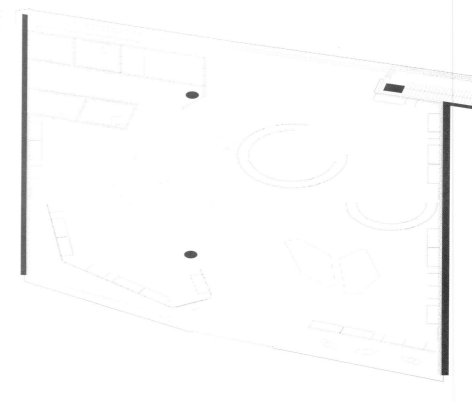

This space, furnished with volumes, creates a fluid ambiance which allows the flow of both light and people, in dialogue with both the inside and the outside of the mall, towards the city.

On the walls the wood covering repeats itself in vertical lines and shadows that we can think as infinitive.

Only people can activate this space following fluid paths which foresee pauses in limited areas dedicated to the dream of cloth.

BAO BAO
ISSEY MIYAKE

Design Agency:
MOMENT Inc.

Design Team:
Hisaaki Hirawata,
Tomohiro Watabe

Client:
ISSEY MIYAKE INC.

Location:
Tokyo, Japan

Area:
20 m²

Photography:
MOMENT Inc.

This shop has two faces, white and black. This two colored triangular pattern rhythmically dances in the shop. It looks that there is no border between the wall and commodity. The proportion of white and black is gently changing on the long wall. This flexibility embodies the versatile possibility of BAOBAO.

This long and narrow shop is 9 meters long and 1.5 meters wide, which is like an alleyway. Although it can not be said that this is a nice location, we positively attempt to make use of this unique condition. As a handbag, Bilbao, is comprised of the repeated triangular pieces, the shop embodies this characteristic by spreading it all over the wall. It has two faces, black and white. The colored pieces look like that it dances in rhythm. A passerby would enjoy the two-sided image. There are only two simple shelves, however the goods are brought out different images because of the background. There is no shelf, hanger rail, or display case, but only a long bench on the other side. The customers would also enjoy graphical display, and be stimulated more interests in the goods. The harmony between goods and background wall hopefully makes a multiplier effect.

ESTNATION Bis

Design Agency:
MOMENT Inc.

Design Team:
Hisaaki Hirawata,
Tomohiro Watabe

Client:
ESTNATION INC.

Location:
Tokyo, Japan

Area:
264 m²

Photography:
Nacasa & Partners

ESTNATION is named under two reasons. One is from "East Nation" because ESTNATION attempts to extend to the world from Japan. Another is from "EN" which means destiny in Japanese. As they sell the western clothes, the store space needs to embody something related to the east country, Japan. The vertical lines are more seen than the horizontal lines in Japanese culture, which is why the vertical design is used for the store. The key design is the white stick-partitions that vaguely divide the whole space. As the location and environment of the store is not said perfect because of the low ceiling and deep shape, The stick-partitions help to show customers the whole of interior even from the entrance. If they were wall partitions, the store would be very close and narrow so that the store might lose customers. The relief customers also work as a guide for them. Beside, the wood material and the indirect lighting are used for the entrance area to show a warmness to welcome the customers. The lacquered rectangular panels as one of the decorations and the shoji paper on the wall flash across our mind as Japanese. It is attempted to create the west and east mixed atmosphere.

SCFashion

Design Agency:
OOBIQ Architects

Location:
Shenyang, China

Client:
SCFashion

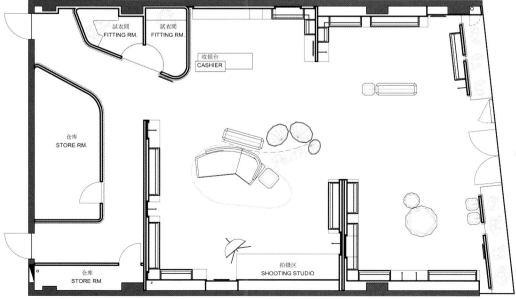

OOBIQ Architects tried to artistically interpret women's most distinctive qualities in its recent interior design project – SCFashion.

The fashion always remains dynamic. The designers practiced the broadest extension of artistic creativity, making this boutique into a fashion set. The sight of the special object, which looks like a diamond as well as a camera, awakens the feminine heart, brings back memories of a noble life and creates a dramatic, dynamic scenario where a beautiful woman is happily posing for photographs.

An interior design echoing the sentiments of every woman – the feminine heart – definitely stirs the women's imagination.

The lines and curves are attempts to shape the interior on the woman's silhouette. Sculptural wall made of translucent materials presents a unique feeling of excitement and a sense of comfort. That's very feminine.

Labels Clothes Shop

Design Agency:
Maurice Mentjens Design

Client:
Nino & Anita Schmeitz

Photography:
Leon Abraas

Steel trees form a reference to the Garden of Eden. White refers to virginal innocence, black is for the lost paradise, a subtle feel for mythology and the mystical is seen in all the work of interior designer Maurice Mentjens. He combined three small spaces and then divided them again into Yin and Yang. Back to the origins of fashion, "the mother of all the arts".

Labels in Sittard offers an impressive collection of youthful, trendy fashion brands. A selection of the brands on offer: Comme des Garçons, Givenchy, Moncler, Rick Owens, Maison Martin Margiela, Y-3, Acne, PRPS, Neil Barrett, Canada Goose, Damir Doma, Dondup, See by Chloe.

The existing shop soon proved to be too small, and was extended last year with the addition of a neighbouring premise. The long, narrow garden in between the buildings was given a glass roof. By breaking through windows and doors in the side walls, three interconnected spaces were created with seven passageways. Mentjens created a division between the women's and men's sections precisely in the middle of the central glass-roofed space, starting with the floor. One half of the floor is pure white, the other jet black, colours representing Yin and Yang, the feminine and the masculine. Elegant versus rugged, like the intangible, graceful world of Venus and the earthy, black smithy of Vulcan, her husband, in Roman mythology.

The connection between the spaces is made by two sales counters, half protruding into the glass-roofed space and half into the white or black areas. Thanks to an ingenious anchoring system in the wall, the blocks are almost magically suspended in the spaces. Only the very ends are subtly supported by a plexiglass foot.

Particular eye catchers in the design are the steel trained trees in the glass-roofed space. They make handy clothes racks, but are primarily a reference to the lost paradise. The former inner courtyard immediately conjured up associations with the Garden of Eden in the mind of Maurice Mentjens. "It's the biblical story of creation, in which Adam and Eve are free of sin in the beginning in the Garden of Eden. They only start to clothe themselves after their Fall, when they had eaten of the forbidden fruits from the tree of knowledge of good and evil, and become aware of their nakedness. They first clothed themselves with fig leaves and later with clothes made for them by God from animal skins. In this regard, you could see the design of clothes as the first creative deed in humanity and thus as the mother of all the arts." So says an impassioned Mentjens, who sees fashion as a fully-fledged art form.

plattegrond

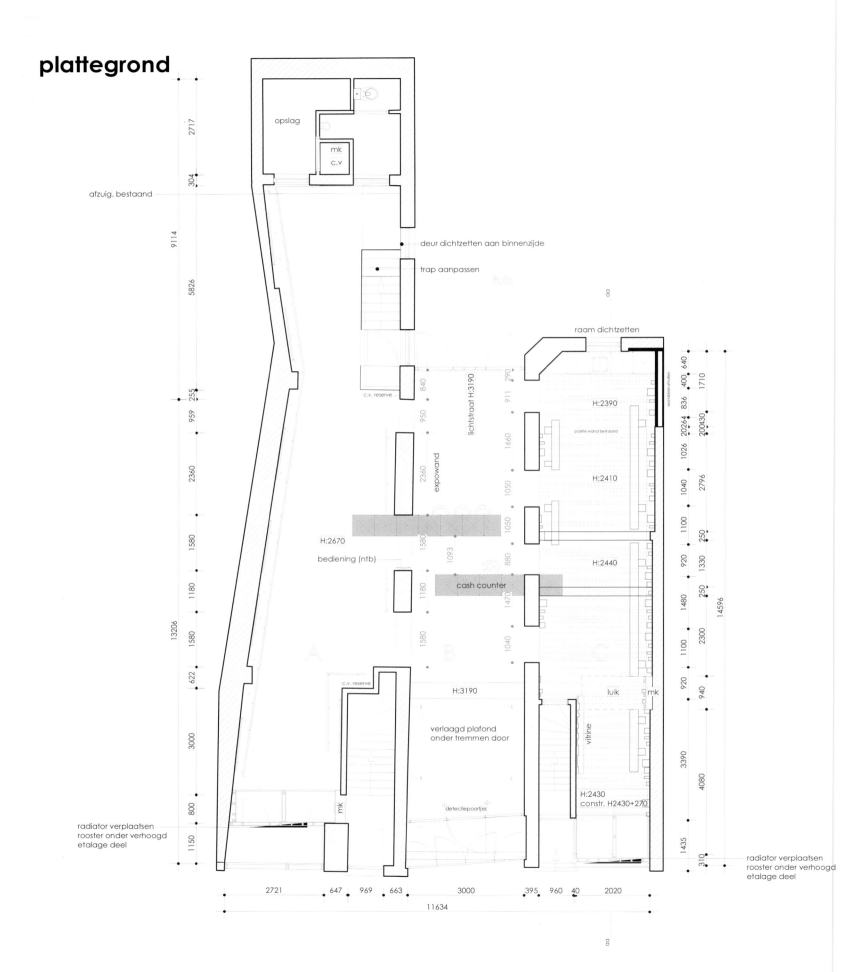

opslag

mk
c.v

afzuig. bestaand

2717

304

9114

deur dichtzetten aan binnenzijde

trap aanpassen

5826

fuin

raam dichtzetten

255

959

c.v. reserve

840

290

lichtstraat H:3190

911

wandstaal afvoeren

400 640

1710

H:2390

20 264

200 430

positie wand bestaand

950

1660

1026

2360

expowand

2360

1050

H:2410

1040

2796

H:2670

1580

1050

1100

bediening (ntb)

1093

880

H:2440

920

250

1330

cash counter

1180

1180

1470

1480

250

14596

1580

1580

1040

1100

2300

622

c.v. reserve

H:3190

luik

mk

920

940

verlaagd plafond
onder tremmen door

3000

vitrine

3390

4080

H:2430
constr. H2430+270

radiator verplaatsen
rooster onder verhoogd
etalage deel

800

mk

detectiepoortjes

1150

1435

310

radiator verplaatsen
rooster onder verhoogd
etalage deel

13206

2721   647   969   663   3000   395   960   40   2020

11634

ESTNATION
Fukuoka

Design Agency:
MOMENT Inc.

Design Team:
Hisaaki Hirawata, Tomohiro Watabe

Client:
ESTNATION INC.

Location:
Fukuoka, Japan

Area:
404 m²

Photography:
Nacasa & Partners

ESTNATION is a fashion shop collecting stylish clothes, shoes, and accessories for ladies and men. The brand is named according to the concept of EAST-NATION, in short, spread from the Eastern countries to the world. Therefore it is important to harmonize the modern design with Eastern work and materials. The layout planning is the core to start design. For example, there are many alleyways in Japanese old towns, and it made the walking all the more attractive. The necessity room as fitting and storage rooms are firstly located around the center of the shop. Then, the space between walls and these rooms becomes like the alleyways. They are actually obstructs as the customers can not look around the whole view of the shop. However this structure would makes them stimulate the curiosity and walk ahead deeper, then they find more goods. It is a Japanese way of thinking not to show whole at once. We can find more if we find little by little. The difference of ceiling and floor level gives them more fun to walk ahead the alleyways. There are wooden sticks along the glass wall, thus sunshine comes into the shop. The design of the sticks is taken from Japanese traditional shape, two long and three short patterns. On the light brown wooden panels inside the shop, the special processed wood panels are mixed. This pattern is made by a chisel, and this is also one of Japanese great techniques. The use of traditional eastern way and modern materials such as silver colored stucco wall shows not only great contrast, but also harmony.

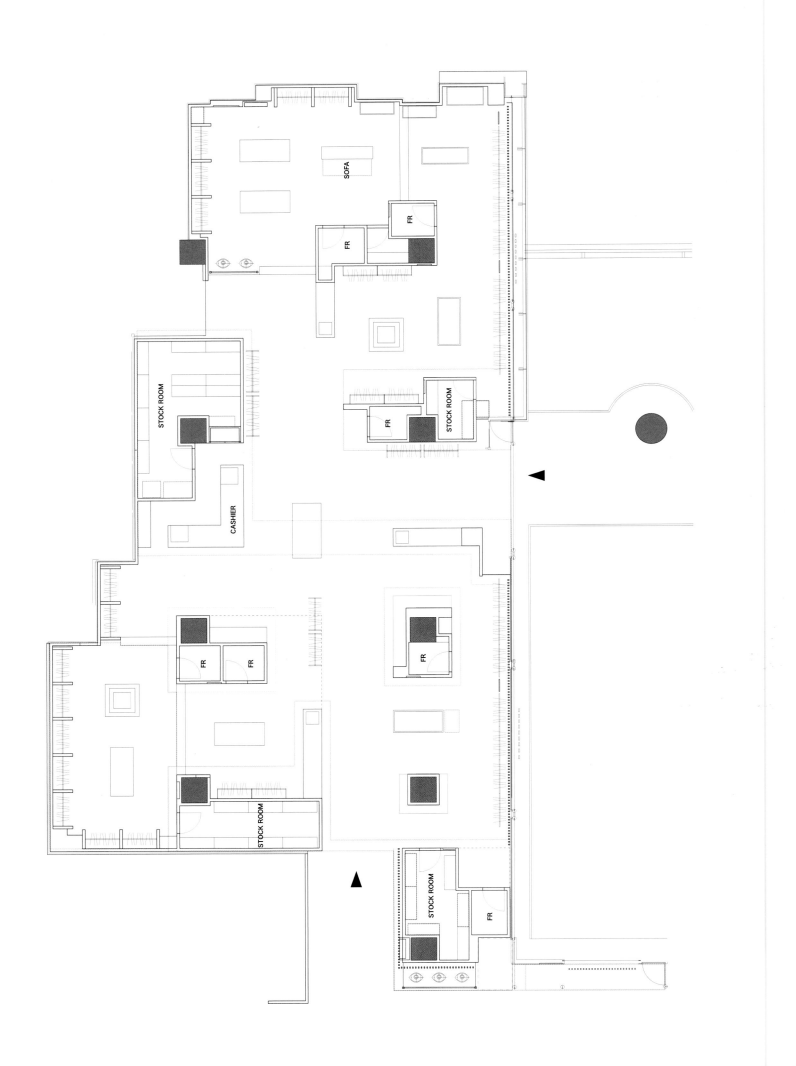

SOFA

FR

FR

STOCK ROOM

FR

STOCK ROOM

CASHIER

FR

FR

FR

STOCK ROOM

STOCK ROOM

FR

LARA
Kanazawa

Design Agency:
sinato Inc.

Client:
CLUB Corporation

Area:
77.32 m²

Designer:
Chikara Ohno

Location:
Kanazawa, Japan

Photography:
Takumi Ota

An apparel store for girls in a shopping mall located in Kanazawa Japan.

The store has very narrow frontage from the passage of the mall. And the store turns to the right at few steps inside from the frontage and has deep space after it.

It's L-shaped site that is difficult to use as an apparel store. The most important thing was to bring people deep inside the store. Designer found the wall which faces to the frontage was comparatively big in the area of this store. So wanted to make the best use of it.

The designer came up with an idea of making the wall with 7 layers, 6 expanded-metals in different mesh, 1 mirrored wall behind the expanded-metals. 3 expanded-metals from surface side (store side) are turned up and become the curving wall which receives people from the frontage.

The expanded-metal seems to be very complicated pattern when it is layered, but when it is turned up and stands as single wall, it loses the presence and changes to just a guideline which leads people inside the store without obstructing the view of the whole interior. The wall with layers also changes its pattern and depth as the surface are turned up and decreasing the number of layers.

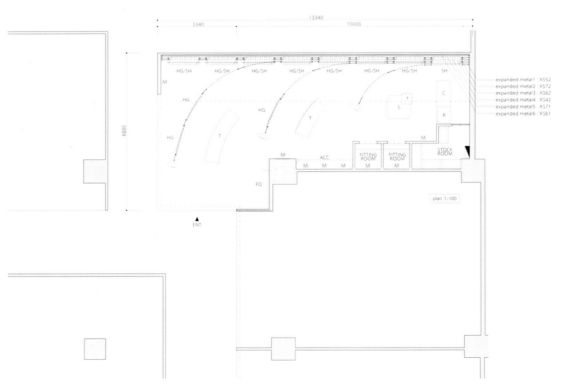

LARA
Kofu

Design Agency:
sinato Inc.

Client:
b.x.store Co., Ltd.

Area:
62.17 m²

Designer:
Chikara Ohno

Location:
Kofu, Japan

Photography:
Takumi Ota

An apparel store for girls in a train station mall located in Kofu Japan.

The site is on the ground floor and has two opposite facades to inside and outside.

Chikara Ohno wanted to make the best use of this condition and he came up with the idea to divide space with lattice plane in order to make horizontal void space where no item interfere the view and connect (tie) two facades. (Below the lattice = shop space / above the lattice = void space)

Its lowest point is 1350 mm from the floor, and it stretches to 3000 mm at its height. People don't hit their head because there is always a big opening at low point above the line of flow.

You can poke your head above the lattice plane at the bottom of the ravine. It makes a kind of blank in the act of shopping and soften the experience of it.

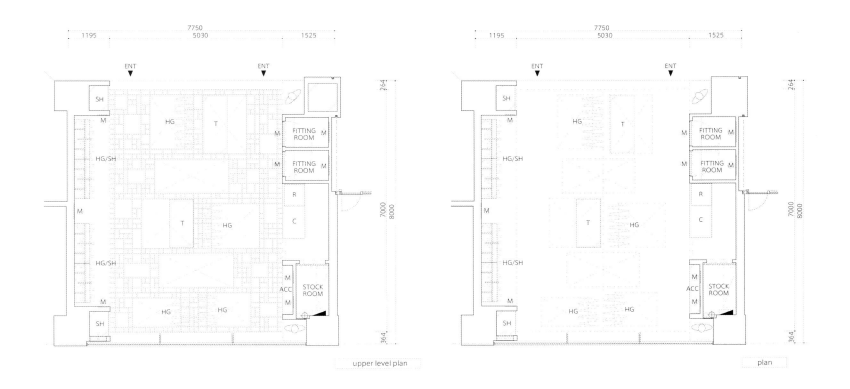

upper level plan

plan

Note et Silence

Design Agency:
Specialnormal Inc.

Designer:
Shin Takahashi

Location:
Hyogo, Japan

Area:
82.36 m²

Photography:
Koichi Torimura

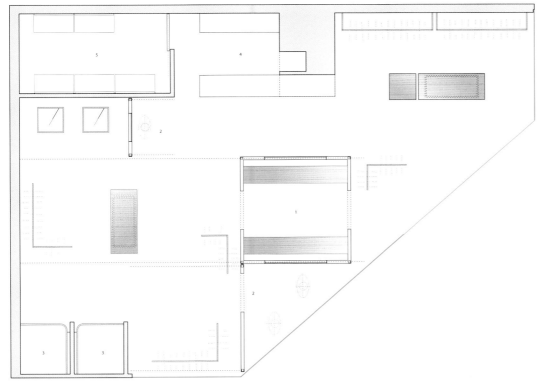

Note et Silence is a select shop, which was newly opened in Kobe, a port city in Kansai area. The shop holds three original brands, which appeal to sophisticated women with a sense of humor and with a playful mind.

"Humor" and "Playfulness" were the keys when Specialnormal started this project, and they defined the shop as a stage, customers as actors / actresses, and the interior as a stage setting.

Based on the idea of "Stage Setting", Specialnormal aimed to create a versatile space and they made the Wall (= Box), which can be moved to a certain extent. If the box is placed near the entrance, it creates a corridor and the space becomes very aggressive atmosphere. If it is placed at the back of the shop, it creates a bigger space within the shop. With the effect of the box, the shop space can be configured for different scenes like a gallery.

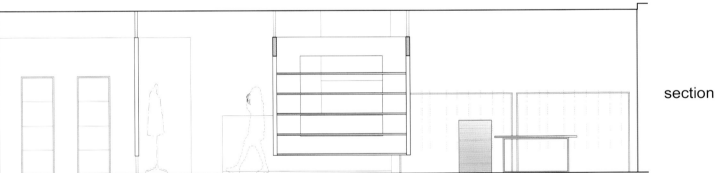

section

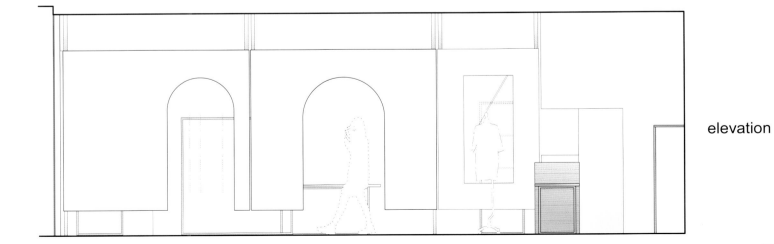

elevation

Fast-shoe Munich

Design Agency:
Bailo+Rull ADD Arquitectura

Client:
Munich Sports

Location:
La Roca del Vallés,
Barcelona, Spain

Area:
160 m²

Photography:
Albert Marin, Pol Viladoms
and Xavi Manosa

The innocent and fresh vision of kids is the leitmotiv of the new shop that Munich Shoemakers have opened in La Roca Village Shopping Complex in Barcelona. From the start, the project uses the imaginary set around the question "How do things work?" commonly asked by us in our childhood.

The shop plays with the idea of moving Munich's factory in Capellades to La Roca. The aim is to make the costumer believe that, behind the eight meters wall, a team of 21st century shoemakers are doing their pair of shoes in the same moment. The shopping experience becomes something more exciting: to buy a Munich's pair means to become, for a while, a lucky kid.

The amazing conveyor belts system invading the ceiling, crossing and rotating when necessary.

The election of materials was easy from the start: the intention was to recreate an industrial atmosphere using the products that the conveyor belt wholesaler offers, and making a reflecting environment that multiplies the belts to infinite. The pavement galvanized steel all the furniture is also done with the conveyor belt rollers.

The project wants to fix the memory of a kid that once visited, with their parents, a shoe shop where a team of shoemakers hidden behind a wall made his first pair of Munich shoes. They let them fall through a conveyor belt to the counter, like if they were "freshly baked".

First Boutique

Design Agency:
Silvia Simionato Architetto

Location:
Padova, Italy

Client:
Calzature First Life srl

Collaborator:
Luciano Bordin

Situated in Padova, Italy, busy central shopping area, the new configuration of this store is very simple.

This explains why therein no tradictional display system as such, which means that the shoes and accessories are instead arranged haphazardly on assorted tables, almost as if left there by accident.

The result is a curious scenario that is both playful and sophisticated, something like a film-set that work principally at psychological level.

Clean lines, classical material; wood, plaster, glass and brass help to create a relaxing atmosphere.

Tea-tables of different sizes allow exposure always different according to the needs and the seasons, the cash is hidden in a private and intimate space.

The shoes are tested in a boudoir elegant and intimate, white moquette blue and glass chandelier, that is not seen when approaching the window.

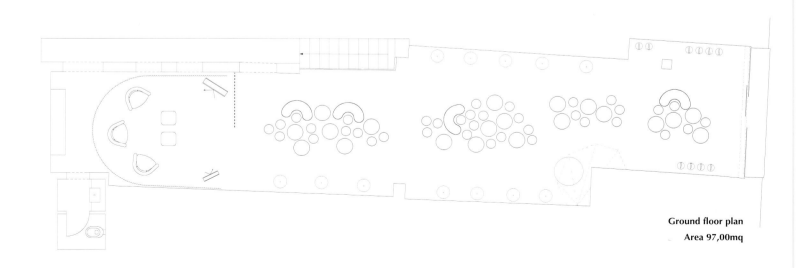

Ground floor plan
Area 97,00mq

Stuart Weitzman Shop

Design Agency:
Fabio Novembre

Design Team:
Lorenzo De Nicola,
Domenico Papetti,
Alessio De Vecchi

Client:
Stuart Weitzman Inc.

Location:
Via dei Condotti n°27,
Rome, Italy

Area:
95 m²

Photography:
Alberto Ferrero

"They searched and poked around my closets, looking for skeletons, but, thank God, all they found were shoes, marvelous shoes," smiled Imelda Marcos, the former First Lady of Philippines.

Imelda Marcos eliminated the feelings of guilt in women who, like Carrie Bradshaw in "Sex and the City", now uninhibitedly flaunt their fetish objects.

Shoes are the perfect modern project: a product of superb craftsmanship transformed from a mere object into a thing of desire.

Clothes do not adapt to changes to the body, but shoes always fit the feet that buy and wear them: perhaps that's why women love Stuart Weitzman.

When Stuart asked Fabio Novembre to design places for displaying the shoes he designs, Fabio Novembre immediately thought of precious boxes wrapped up, "Christo style", like gift boxes for his customers, with an uninterrupted ribbon that runs all along the space, creating paths on which the shoes trace vectors of desire.

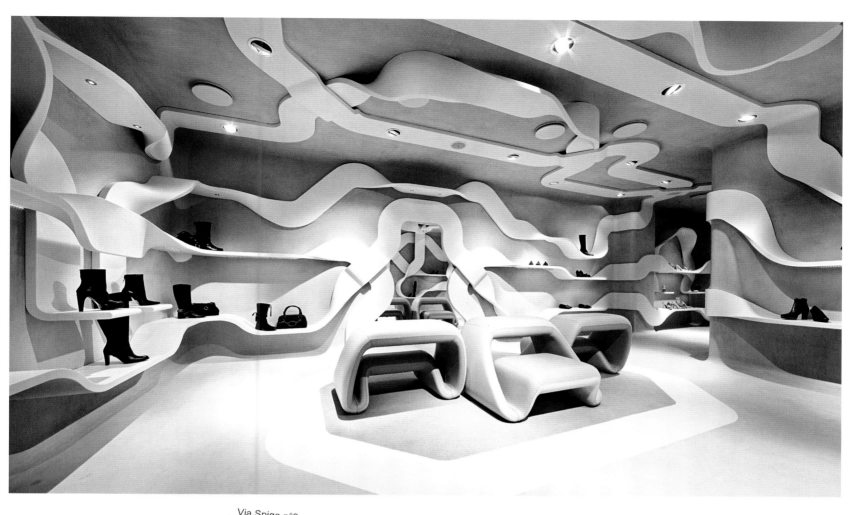

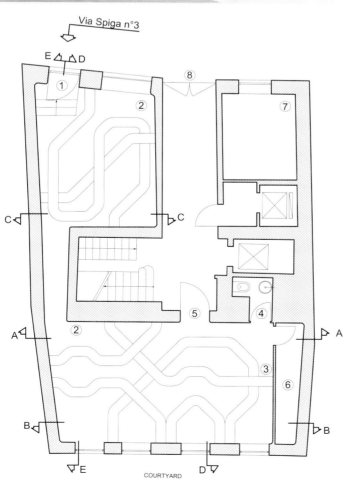

Via Spiga n°3

E D

C C

A A

B B

E D

COURTYARD

GROUND FLOOR PLAN

1 main entrance
2 shop area
3 cash area
4 service toilette
5 back door
6 storage
7 shop window area
8 building gate

Harvey Nichols

Design Agency:
Four-by-Two

Location:
Edinburgh, UK

Harvey Nichols launches its all-new luxury "shoe boutique" department in Edinburgh, created by international retail design agency, Four-by-Two.

The 1300sq ft. haven for footwear is now sited in what used to be the store's hair salon. It's a striking space that stops people in their tracks and welcomes them in to experience the world of beautiful footwear where they can try on luxurious brands such as: Alexander McQueen, Christian Louboutin, Saint Laurent, Gucci and Jimmy Choo.

The brief from Harvey Nichols was to create an environment that fits with the existing architectural language of the department store, yet, importantly, feels like a distinct destination, while still flowing seamlessly from other merchandised areas of the store. Additional stock space needed to be considered and customers had to be able to try on their shoes in ultimate comfort, able to reach a mirror without having to walk on hard surfaces.

Above all, the boutique needed to create intrigue by using unusual and rich textures, materials, shapes and lighting that would delight its customers, whilst at the same time ensuring that the shoe itself was the hero.

Four-by-Two responded to the brief by creating a department that uses the full depth of the available footprint, with concealed stockrooms at each end balancing the long slim proportions of the total space. The boutique's perimeter has been designed with "open arms", with wide sight lines encouraging footfall as customers are subliminally drawn into the space.

Angled walls play with perspective and are designed to look singular and flowing, allowing the subtly changing directions to create "zones" for the range of luxury shoe brands. These zones give individual brands the ability to grow and contract with seasonal product change, without having to endure a regimented "bay" look.

New floor finishes were installed to match the existing store limestone. Luxurious marble shelves are fixed to walls; each individually shaped and designed to "float" in the space. Feature merchandising tables invite customers to interact with the product on "fanned" marble disks. Each table is dramatically lit from above with narrow beam-spots and striking gold pendant lights set in a circle.

Sumptuous circular seating covered in rich gold velvet allows customers an element of "privacy", as they face away from each other when trying on their chosen shoes. The banquets are highlighted with oversized gold lighting drums with soft diffused shades set at an angle from above. A rich muted colour palette of deep plush bespoke designed wool rugs sit under the ceiling drums and seamlessly take the customers to the mirrors. These mirrors are carefully designed into each corner of the boutique, deepening the perceived depth of the space whilst allowing customers a 360 view of themselves. Existing columns were then clad in jewel shaped glass, mirroring views of the overhead gold pendants and sending reflections around the store to attract attention.

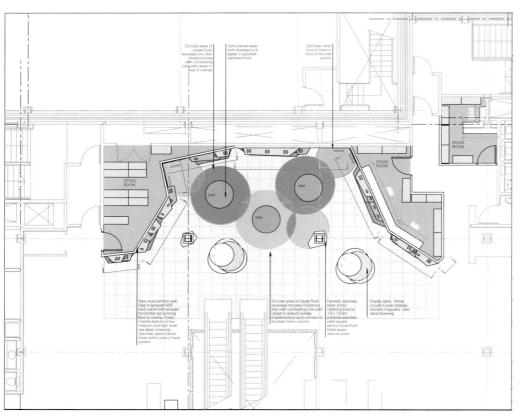

Circular areas of
carpet flush
recessed into new
limestone tiles
with contrasting
coloured carpet in
area of overlap

Upholstered seats
with recessed kick
plates in polished
stainless finish

Opticlear silver
mirrors fitted in
front of the wall
panels

New stud partition wall.
Clad in sprayed MDF
back panel with sprayed
horizontal ribs running
floor to ceiling. Fixed
marble shelves at low,
medium and high level-
see detail drawings.
Opticlear glass shelves
fitted within zone of track
system

Circular areas of carpet flush
recessed into new limestone
tiles with contrasting coloured
carpet in area of overlap.
Carpet to run up to corners of
faceted mirror column.

Faceted, opticlear,
silver mirror
cladding column,
12 x 12mm
polished stainless
steel square
section to be flush
fitted down
vertical joints.

Display table. White,
circular Corian shelves,
stacked irregularly. See
detail drawing

Minna Parikka
Flagship Store

Design Agency:
Joanna Laajisto Creative Studio

Photography:
Katri Kapanen

Location:
Helsinki, Finland

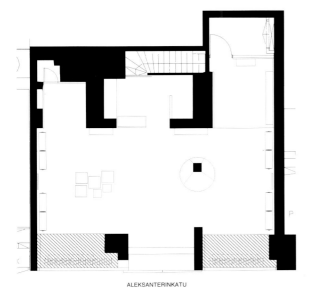

ALEKSANTERINKATU

The shop is located on Helsinki´s main shopping street Aleksanterinkatu, with large storefront windows opening out to the busy walkway. The interior of the store is simultaneously sophisticated and playful. The designer Joanna´s design concept is inspired by Parikka´s feminine shoe and accessory collections.

Classic natural materials, such as leather, wood and marble, work well with the pastel colored background. The walls and glossy epoxy floors are a shade of faint powdery peach while the display shelves have mint and light coral hues. The color palette is muted so the merchandise gets the main attention. The use of mirror in large quantities gives the space some extra edge.

Although the interior is elegant, the store doesn't take itself too seriously. Fun details, such as the "kissing booth" with tan leather lip sofa and neon xxx sign can be found in the floor to ceiling mirrored corner of the store.

Joanna designed a delicate white marble writing desk with neon pink thumb stone style lettering to serve the purpose of a cashiers counter. The company slogan "May these shoes lead you to new adventures" sums up the thought behind the brand. Minna Parikka's shoes and accessories radiate playfulness that takes women on their own individual adventures.

The Helsinki-based label has stockists in over 20 countries. Since 2006, Minna Parikka's shoes and accessories have graced the feet and bodies of women all over the world, including Fergie, Paloma Faith, Beth Ditto from the Gossip and Dita von Teese. Lady Gaga is true fan: Minna has designed a double platinum award for her and in Autumn 2012 Terry Richardson featured Gaga wearing Minna's limited edition cartoon shoes in his blog.

Although the brand´s main focus is footwear, buttery soft leather gloves, bags, wallets and surreal silk scarves complement the range.

MAY THESE SHOES LEAD YOU TO NEW ADVENTURES

MiniMunich Store

Design Agency:
Dear Design & Architecture

Design Team:
Ignasi Llauradó, Eric Dufourd

Client:
Munich

Location:
La Roca Village,
Barcelona, Spain

Area:
35 m²

Photography:
Xavi Mañosa

MUNICH, Barcelona's Brand for fashion footwear and sports reopened its children boutique MINI MUNICH in La Roca Village, Barcelona.

After the success of its first children's shop in this mall, MUNICH opens a new space dedicated to the smallest in another location with a daring and innovative design.

This project, created and designed by Dear Design, is premised on the characteristic imagination and curiosity of children to create a scenario according to the way they experience and interact with the space.

The aim of the project is to transport them to a new imaginary world that may be explored.

The main idea was to reinterpret a module able to transport anyone entering the store inside a submarine or a spacecraft, to enter a new world. The simulated hermetic gate door of the store reveals from the beginning the feeling of getting into a new experience.

Thus, in this new 35m² location, children can enjoy and leave their imagination fly, making go shopping a pleasant and fun time.

This time the concept breaks completely with the previous MINI MUNICH and creates a new scene without losing the daring spirit that characterizes the brand spaces.

In this MUNICH new store, the space with white walls and glazed lit plates scattered over the shelves, add character and create a world of science fiction. On its walls and ceiling appears real magic: oval slits that reveal synchronized images of distant outer space or the depths of the sea.

All this makes this new MINI MUNICH boutique a versatile space that will make each purchase a new experience for both adults and children.

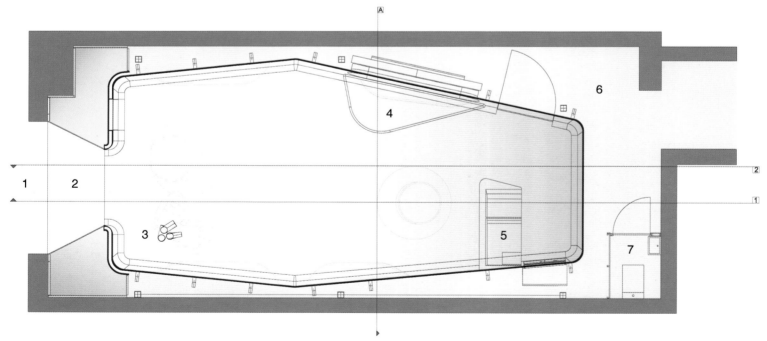

1 Gate door
2 Mirrors filter
3 Periscopes
4 Pouffe
5 Cash desk
6 Warehouse
7 WC

M Dreams –
Melissa Flagship
Store

Design Agency:
Blu Water Studio

Design Team:
Siew Hong Lai,
Elaine Yap,
Bernice Chan

Client:
Ominiscent Sdn Bhd

Location:
One Utama Shopping Centre,
Pelating Jaya, Selangor, Malaysia

Area:
121 m²

Photography:
Lin Ho

The Melissa brand – known internationally for their plastic fantastic shoes, chose Blu Water Studio to design their Malaysian flagship store in One Utama Shopping Centre (Petaling Jaya, Selangor). The award winning interior design studio was inspired by the manipulation of sheet of paper to create the look for the store.

A singular M Dreams logo in darker color adorns the entrance, giving the store a subtle signage for the brand. In order to achieve a mold to suit Melissa's flexible shoes, Blu Water explored the shapes and outcome of compressing sheets and peeling layers. The peeling effect is apparent on the right side of the shop where, make-believe layers of sheets are peeled off the wall forming shelving for the display of the brand's colorful product. On the opposite side, giant fingers can be imagined pressing and pinching the floor and wall, fashioning ribbons out of them to create the rows of multi level cluster displays. A curtain shape wall hangs from the ceiling that runs across the peeled off shelves to the cashier area feature projections that promotes Melissa's ongoing campaigns. The overall whiteness is the one and only features uniting Melissa stores all over the world.

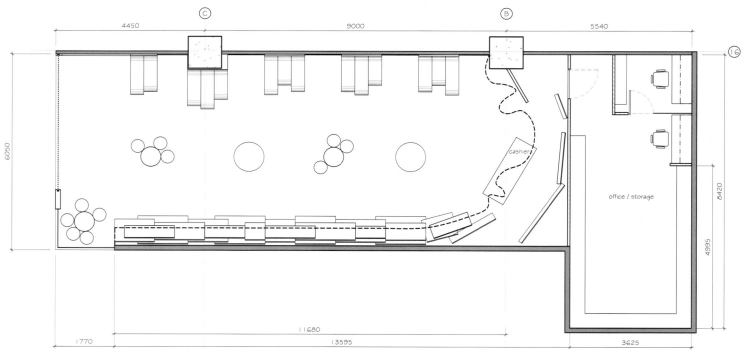

CAMPER STORE
Washington

Design Agency:
Miralles Tagliabue EMBT

Designer:
Benedetta Tagliabue

Client:
Camper

Location:
Washington, USA

When Miralles Tagliabue EMBT was asked to designed a store they started to dream Camper has much to do with "el campo", that is the countryside, the fields, to walk in the fields, and so they imagined shoes stepping on irregular surfaces, like when they walk on earth.

Camper has gotten them used to play, smile, see fashion with children's eyes, they imagined trying on shoes in front of the twisted mirrors of an amusement park, here Benedetta Tagliabue is tall tall… here short short… and that's how wobbly reflecting surfaces appeared in the project.

Later, EMBT visited Camper´s factory to see how shoes are made. There they understood they wanted to build their new store the same way Camper shoes are built in Mallorca: flat leather sheets are cut and, like magic, with a couple of stitches and a mould, they become beautiful tridimensional wrappings!!

EMBT cut different shoe forms, with hills, without hills, tall, short, for men, for women, they put those flat profiles one next to the other, and that is how their the volume of their new Camper store appeared!!

Its construction was unusual, but very simple (Benedetta Tagliabue loves thinking how much fun the guys who assembled it must have had!) MDF sheets cut like shoes, fixed one next to each other transform into benches, tables, surfaces and lots of crazy mirrors – where children will spend their time while their moms devote time to the art of shopping.

Ferrari Store
New York

Design Agency:
Iosa Ghini Associates

Designer:
Massimo Iosa Ghini

Client:
Ferrari Spa

Location:
New York, USA

Area:
192 m^2

Entirely created by Studio Iosa Ghini, the Store is laid out on an area of approximately 186 m^2 and fronts directly onto Park Avenue: a wide glass facade enhanced by video projections and graphic panels, nonetheless making it possible to view the inside of the store. The sales area, arranged on a single floor, extends inwards with the dynamic and sinuous lines inspired by the Ferrari automobiles. The dynamism and idea of speed are underlined in this project by red and black "ribbons" that move inside the space accompanying the visitors as they make their way among the various merchandise areas, all the while helping define the Store area and distinguishing it from that of the Dealership.

The side that fronts onto Park Avenue is characterized by displays at a low height which can be viewed both from the window as well as from the inside of the store. The area dedicated to the Lifestyle collection continues along the end wall, culminating in the sculpture-like payment desk. The payment desk has been researched to ensure a highly visible position, separating itself from the luxury wall by means of a red ribbon that accompanies the whole path of the visitors through the Store. Thereafter, the furnishings develop into a slat furnishing system, studied with care being given to the details: aluminum slats combined with various accessories allow for particularly flexible display configurations. The central area is characterized by light boxes and a display shelf where a case dedicated to memorabilia is located.

Camper Store

Design Agency:
A-cero Joaquín Torres & Rafael
Llamazares Architects

Location:
C/ Mesones 51 bajo, Granada, Spain

Area:
48 m²

Construction:
Obras y Otras cosas S.L.

Photography:
Juan Sánchez

On Mesones street, one of the best known, commercial and important streets of Granada, Spain, there's a little store of 48 m² whose reform has been developed by A-cero. The customer is the Spanish brand Camper, dedicated to the exposition and sale of shoes. It will be its first store in Granada.

The first important point is the use of the space, because a wide and useful place is needed. A-cero was selected for this work because of its actual, fresh and dynamic design, typical in Joaquín Torres and Rafael Llamazares' projects. Each store that Camper opens is an event, depending, among other things, on the designer chosen for the project, because each one has his own style. Camper always looks for a new style and concepts for their stores, making different and special places.

The store is well-located in the most famous and commercial street of Granada. The store, with 48 m², was a clothing shop partially abandoned.

Two colors: white and red, typical of the A-cero interior design projects and Camper. The space is opened and projected with organic elements like the expositors, made of lacquered wood with white shelves and red face, having an indirect lighting by leds. Also, following the floor shapes, there's a red and white bolon. Adding more sculptural values to the complex, in the middle part there's another curve module used as a brech-change. To achieve more depth, the rear wall is a mirror. There is still space for a warehouse at the back. The front is made of composite panel with red aluminium and the showcase of glass and red vinyl following the style of the store.

The final result is an interesting fusion between the corporative resources and designs of Camper and A-cero. It's a special place that improves the market of the area and also, the possibility of feeling the universes of Camper and A-cero.

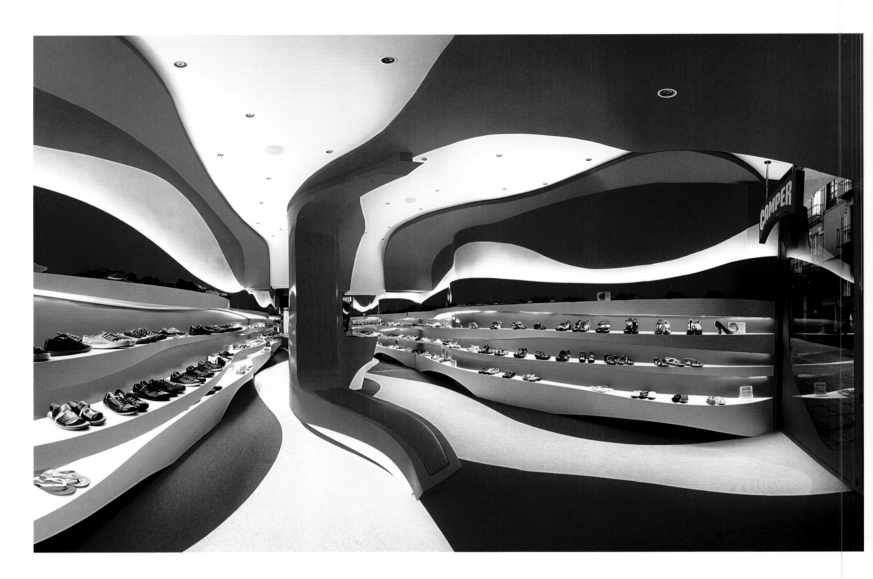

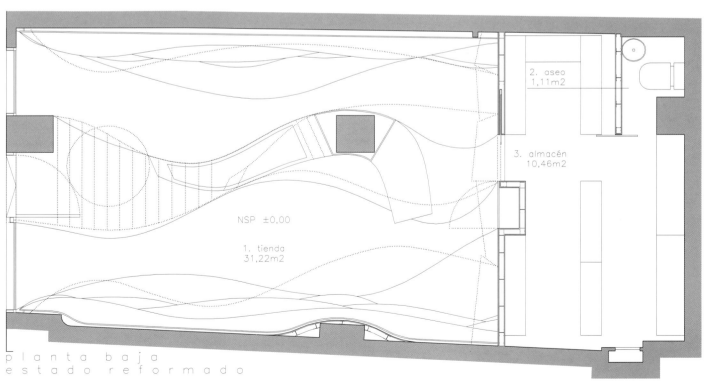

2. aseo
1,11m2

3. almacén
10,46m2

NSP ±0,00

1. tienda
31,22m2

planta baja
estado reformado

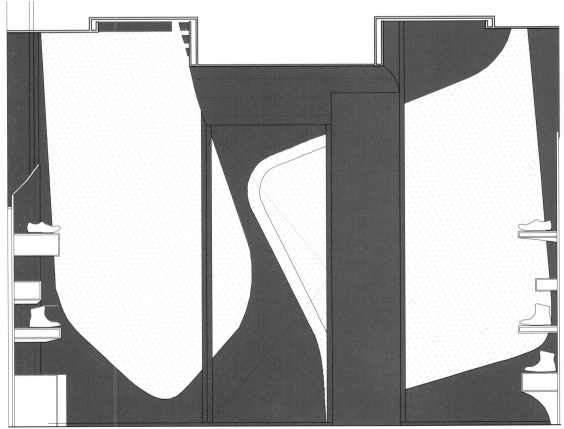

Puma Brand Store

Design Agency:
plajer & franz studio

Client:
PUMA AG

Location:
Osaka, Japan

Area:
579 m²

Photography:
Hiroyuki Orihara

Puma's premium store in Osaka is a new landmark on the retail horizon. Both the exterior and the interior architecture of the space have been pumarized and speak a homogenous language that unifies the company's philosophy, an intelligent yet simple retail design, and the joyful and witty spirit of the sport-lifestyle brand.

plajer & franz studio under the direction of Ales Kernjak (Head of global store concepts) developed a retail concept that is simple and flexible while integrating local references and offering Puma's clients a joyful shopping experience.

Thereby, the new premium brand store is more than just a shopping place. It is a social and cultural meeting point. It's a space for events and happenings of various kinds. The lower two floors of the four-storey Puma building have been designated for the 600 m² shopping space, while the upper level – an open roof top, surrounded just by a light façade construction – creates an open space for performances and sport events.

The façade is light, made out of meshed metal baffle allowing daylight to enter the store. At night, the interior lighting beams out, illuminating the street and giving a sneak preview of the shop's inside during closing hours. The interior design unites the individual concepts for sportlifestyle (including motorsport), performance, and black label. It aims at strengthening the brand image, enhancing the display of Puma's collections, while celebrating footwear as the core of the brand.

An impressive, cone-shaped staircase in the centre of the store with a red brand-wall in its back, is the eye-catcher making a strong brand statement. It draws attention not only to its design but above all to the footwear tribune at the foot of it. The "stage" continues as a footwear catwalk stretching across the entire ground floor, which is also home to the sport-lifestyle product line and black label. Although the design of both areas is simple and functional with references to Japanese architecture (e.g. origami seating elements in the sport-lifestyle le area or the use of materials such as wood and concrete), both speak different visual languages and are clearly distinct from each other. with a very light and flexible design, the use of materials such as black steel, plywood panels combined with re-used gym flooring elements and the mixture of matte and high gloss surfaces, the black label area carries a visible sports heritage.

The first floor is entirely dedicated to the performance area. Here, the nature of performance is visible and perceptible throughout. The design for this area is even more flexible and functional, using more technical materials such as stainless steel or perforated metal in silver grey. The flexible approach is further reflected in the design of the display tables, which are covered with convertible materials. A catwalk area runs across the entire performance floor with a stage in the centre of it and leading towards the footwear focus wall.

Generally, shopping at the new premium store is above all an experience and in line with Puma's philosophy and an active interaction with the brand. Whether creating their own customized sneakers at the Puma factory, engaging with Puma's sustainable actions summarized in the sustainability journey or enjoying the online experience via iPads, clients shall be surprised, entertained and given opportunities to identify with the brand.

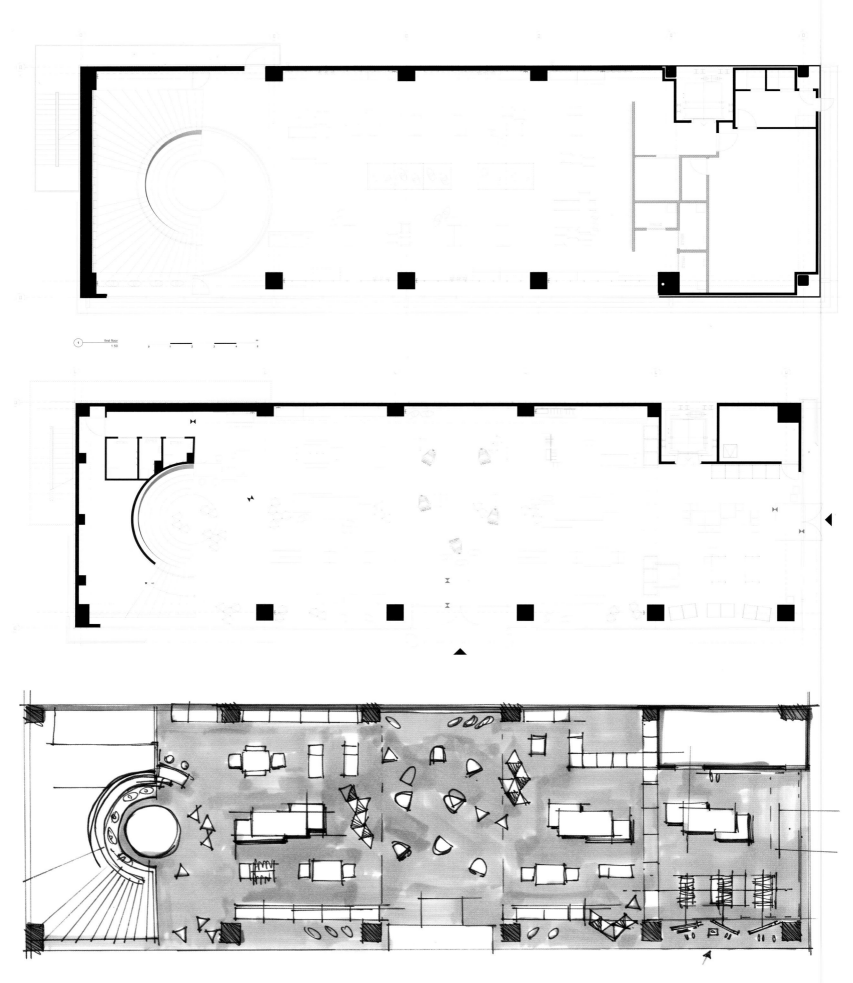

first floor
1:50

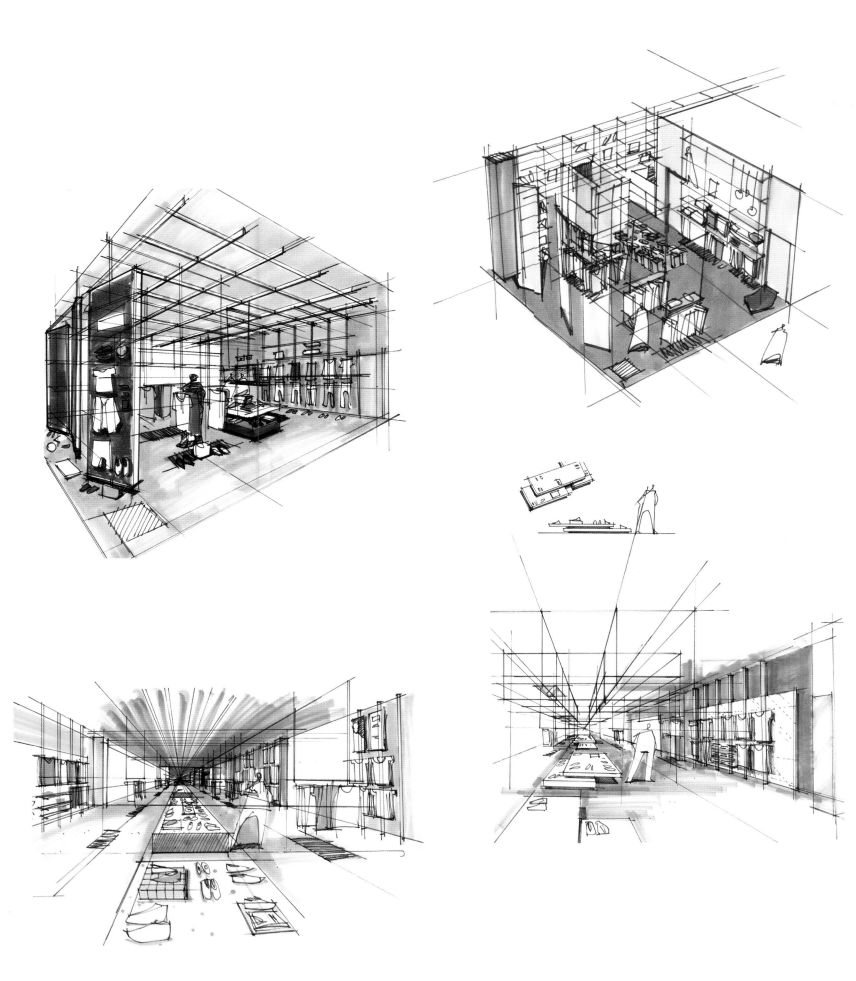

Li Ning Flagship Stores Beijing & Shanghai

Design Agency:
Storeage

Design Team:
Leendert Tange, partner and co-founder,
Storeage Li Kang, Account Director Asia Pacific,
Storeage Ian Clarke, Designer, Storeage

Client:
Li Ning

Location:
Beijing & Shanghai
in China

Photography:
Colin Jones

Storeage, the Amsterdam-based retail design agency, has created a new retail concept for Li Ning's flagship stores in Beijing and Shanghai. Li Ning is one of China's largest developers and distributers of sports apparel, footwear, equipment and accessories.

The two Li Ning stores together comprise more than 6,000 m². The 2,000 m² Beijing store opened in March, and Shanghai, at 4,000 m², opened in July. Storeage started its design process with an analysis of Chinese 18-35s. Findings revealed that 93% view mystique is an important factor in making a brand "cool," and 85% identify with a brand that represents motion.

Based on this, Storeage created a brand experience that celebrates motion and creates a mystique around the idea of movement itself. To capture this mindset, Storeage introduced a strong angular grid pattern that is present throughout both stores – in the ceilings, lighting and floor patterns. This pattern creates an impression of no-stop movement with no beginning or end.

At the Beijing flagship, the theme of motion is at its most visible in the store's façade. When passing by the store, the sweeping lines of the façade's lamels create a continuous sense of movement. At the entrance, Storeage created a massive four-storey void space that evokes vertical movement, like the leap of a basketball player.

The Beijing store's staircase features wavy bamboo slats and a red line that winds from bottom to top, and the ceiling is decorated with curved aluminum lines. At the five-storey Shanghai flagship, Storeage created the same strong sense of movement around the escalator area located at the center of the store. A metal framework with angular details surrounds the escalators on each floor – creating a sense of circular and vertical movement – and is used to display footwear relevant to each sports category.

Munich Kildare Flagship Store

Design Agency:
Dear Design & Architecture

Client:
Munich

Location:
Kildare, Ireland

Area:
60 m²

Photography:
David Murphy

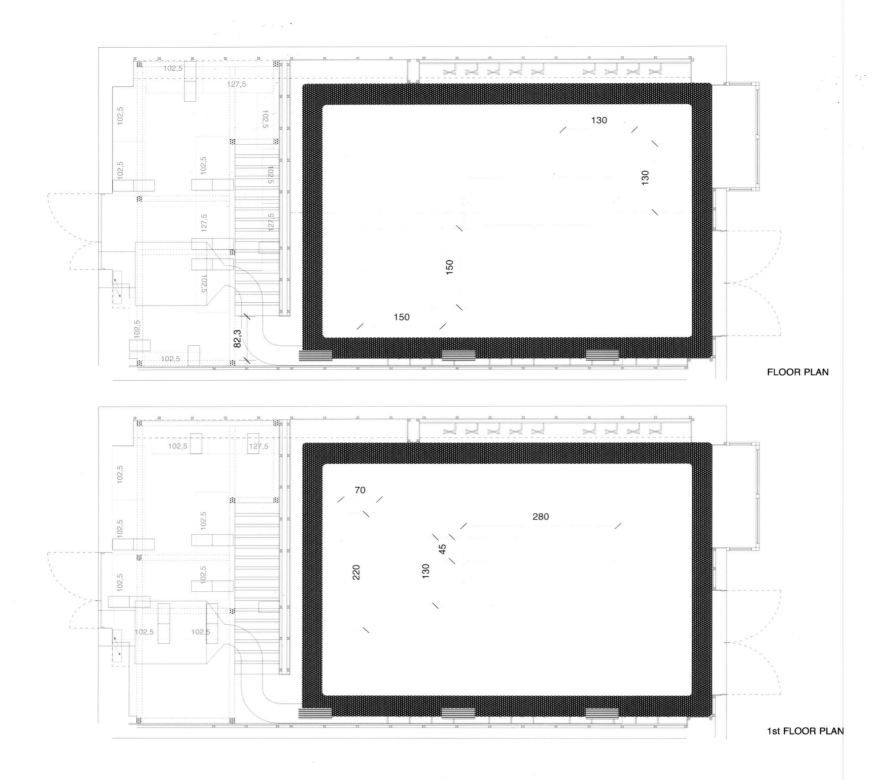

FLOOR PLAN

1st FLOOR PLAN

Munich, the barcelona based fashion – and sports footwear brand opens shop in Ireland. Continuing with its expansion plan, the brand has chosen this country to make its international entrance with its first own shop in the shopping mall kildare village.

Dear design created this 60m² space following the design spirit that characterizes all munich stores: art and design in every shop. The interior evokes a warm and comfortable feeling to make the client feel at home. To achieve this effect, simple and everyday materials, such as the fireproof munich shoelaces, have been used. More than 67km of red laces cover and furnish the room. Red like a roof and dropping like pendulums, the laces spread over the entire ceiling.

To complete the clean design of forms and materials, padded green fabric covers the counter, shelves, sofa and walls, recalling the green hills and mountains of Ireland. This unique design is combined with led-based lighting, chosen to reduce energy consumption, conform to the brands commitment to the environment.

apple & pie Children-shoe Boutique

Design Agency:
Stefano Tordiglione
Design Ltd.

Chief Designer:
Stefano Tordiglione

Location:
One Island South,
Hong Kong, China

Area:
85 m²

Furniture:
Kartell

Photography:
Stefano Tordiglione
Design Ltd.

Playfulness, elegance and practicality come together at One Island South's new apple & pie boutique, the latest creation from Hong Kong-based Italian interior design and architecture firm Stefano Tordiglione Design.

Walking into children's shoe store apple & pie is like entering another world – the giant apple that crowns the doorway hinting at the many delights that lie within. Inspired by the ethos behind the brand's name and its half apple-half pie logo, Stefano Tordiglione Design's concept combines the wellbeing elements represented by the apple with the more playful pie. The former is reflected in the use of environmentally and child-friendly materials with a focus on wood as opposed to plastic for the furnishings, while the latter can be seen in the whimsical interior design which ranges from bright red apple-shaped sofas to imaginative wall displays. There is also practicality behind each well-thought out element. The seating hides storage space, while a lively tree design on one wall with white and red apples hanging from its branches, and elsewhere pie-shaped lattices and mounted fruit palettes, offer ideal shelving opportunities. In the windows, semi-circular pie-like features bring the logo and brand name full circle while providing window-shoppers with a taste of the various European shoe brands that can be found within.

Throughout the space Stefano Tordiglione Design has considered the experience of its clientele, young and old. For children coming to try and buy shoes, the back of the store offers a table at which they can sit and play between fittings, not far from a wall that provides familiarity through its giant blackboard design. Yet the focus is not solely on a positive experience for children. The iconic Kartell chairs surrounding the low table, the Ethel lighting hanging from the ceiling above, and the Giant Red Lamp designed by Anglepoise are design features which lend a sophisticated and elegant air to the store. Coupled with a combination of smooth curves and clean lines, above a warm wood-lined floor and with a colour palette that blends bold reds and vivid greens with a calming mint and light beige, the store effectively and effortlessly moves between the distinct worlds of parents and children.

The flagship apple & pie store sets a precedent – it is a place where children can enjoy choosing shoes that are displayed in a fun and enticing way in cheerful, relaxed surroundings. Through the fruits of its labour, Stefano Tordiglione Design has created the ideal environment in which children's shoe shopping needs can be met.

Lululemon Yorkdale Store

Design Agency:
Quadrangle Architects

Interior Design:
Brothers Dressler;
Lululemon Interiors Department

Location:
Toronto, Canada

Media Support:
Medialab

Contractor:
Summit Brooke

Photography:
Bob Gundu

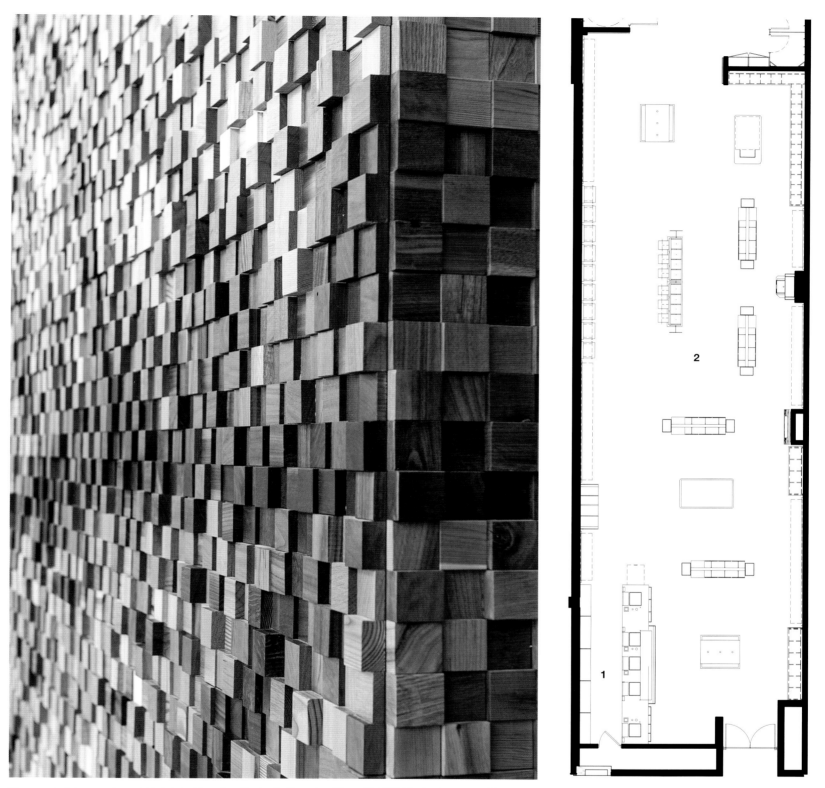

The new Lululemon store at Yorkdale Shopping Centre in Toronto utilizes 35,000+ blocks from over 20 species of wood, and the help of a collaborative team and a beaver to establish a unique retail experience. Created by The Brothers Dressler, Quadrangle Architects and the Lululemon interiors team, the new 279m² Yorkdale Lululemon store is distinguished by its facade with a 7m high pixelated image of a fallen leaf, enticing shoppers with its dynamism, originality, tactility and even its scent. Its warm and welcoming qualities underscore Lululemon's brand affiliation with yoga, harmony and balance.

The facade mixes reclaimed woods salvaged from The Brothers Dressler's furniture workshop that span centuries including the staves of a tanning barrel, shipping pallets, a demolished school and submerged wharf beams, and off-cuts from countless furniture pieces. These woods are now united in a pallet of 29 natural colours to begin a new chapter and purpose. The unusual facade invites shoppers to explore its surface, and as the blocks wrap around the entry, it draws people into the store.

Custom-made pieces by the Brothers Dressler dot the interior: A chandelier made from steam-bent hardwoods mixed with beaver-gnawed sticks from Lake Manitouwabing and embellished by swarovski elements crystals, gives opulence and sparkle to the dressing area, a water station made from veneer mill ends of a walnut tree offers water from a mirrored crevice replenishes fluids, and over the cash desk, a museum artifact style sculpture of an uprooted tree is re-created out of misplaced limbs and sections of found trees and divided into seven parts to represent the chakras of the body. It "breathes" with softly pulsing LEDs, glowing through cast glass sections of bark, exploring the relationship between natural growth, our bodies and the calming nature of trees.

Contributors

A-cero Joaquín Torres & Rafael Llamazares Architects

A-cero Joaquín Torres & Rafael Llamazares Architects is an architecture firm founded in 1996, vowed to the integral development of architectural, interior design and urban planning projects.

The studio has a team of more than 100 specialized professionals, led by architect Joaquín Torres Vérez, working in two main headquarters in Spain, located in Madrid and A Coruña, and a foreign projects office in Dubai.

The practice has evolved in parallel to its clients' demands, building a large portfolio that comprises residential estates, resorts, office buildings, mixed-use high-rises, corporate and interior design. They have the capacity to carry out an architecture project in all its stages, since its conception until the construction management.

Their working method is based on a detailed analysis of the client's needs and the project's program. The result is examined in all its dimensions in order to find every potential problem, all the possible solutions are studied and all the material possibilities for the new building analyzed during the design stage. The concept ideas are kept from the beginning to the end, as a conducting thread for the design process, in order to achieve the best possible results.

Passionate and motivated, A-cero designers develop innovative and visionary designs for their clients. Driven by their core values of individuality, respect, passion and integrity, their commitment to design excellence is integral to everything they do. At A-cero they want to experience the pride and excitement of creation and it is their commitment to be the very best that drives them forward.

Agi Kuczyńska / Inside/Outside

Agi Kuczyńska is an architect and designer from Warsaw. She graduated in 2005 in Architecture and Urbanism from Warsaw University of Technology. She also studied Architecture and Industrial Design at Politecnico in Milan, where after graduating she worked as an assistant at the Faculty of Architecture. After working and studying for many years in Milan and Stockholm, she came back to her hometown. In 2010 she established a design studio Inside\Outside, working mostly with interior design, identification systems and visual identity. Worked with Royal Park and National Museum in Warsaw as well as young polish fashion brands like Zuo Corp and Bynamesakke. Recently she came back to her interests in fashion and industrial design with her friend Kasia Orłowska founded a jewelry brand TAKK launching a collection of customizable jewelry.

AS Design Service Limited

AS represent aesthetic & stylish. It's an aesthetic team created stylish design works.

With vast imagination, great market sense and determination to work together and achieve results with clients, AS design has grown into one of the most anticipated upcoming design firms in Hong Kong. AS design and its team are expertises in providing professional and comprehensive design services to international brand labels in Hong Kong & Mainland China as well as south pacific region.

"AS Founder Four Lau & Sam Sum like the space to fill with the experimental art, and testing the feasibility of commercial space combined with art."

Their strength lies in images renovation for business and corporate, with an aim to assisting brands to stand out and become globally renowned. By providing an appropriate yet high-end image to their target clients, their creative sense will certainly help bring their client's business to a next level.

Bailo+Rull ADD Arquitectura

The approach of Bailo+Rull ADD Arquitectura is to bring the architecture close to the world of processes. This approach is particularly informal because when planning the projects, the emphasis is more the intermittent value of the transient phase than the final value of the finished objects. The architecture is as a never-ending process with committed and transient result, which mixes the randomness with sound and intelligible reasoning. It is a way to produce an architecture where the continuous inestability is considered as an enriching factor that improves the arrhythmic of the project.

The transient architecture has its own way of understanding its relationship with the time and the space in which it occurs. Paradoxically, the awareness of its own temporality shares without complexes its time and space conditions resulting accomplice architecture.

Bailo+Rull ADD trys to crystallize a process of investigation that looks for the balance between assimilation of new ways of reading landscape taken from extra-architectural disciplines and their implementation through a manufacture technology.

Blu Water Studio

Blu Water Studio is an award winning design studio providing interior design consultancy services specializing in hotels, resorts and restaurants. Formed by highly talented designers, the team brings 20 years of experience in the design industry to create compelling and all-encompassing branded solutions.

The studio goal is to create unique and distinctive designs – strive s for passion and individualism at the core of each project, bringing the clientele instant recognition while maximizing the consumer's experience.

The approach is to understand every aspect of our clients' brand and spirit, translating corporate identities into physical forms.

By blending experience, innovation and solid business principles, Blu Water Studio creates memorable and inspiring experiences by architecting outstanding environments.

Chute Gerdeman

Chute Gerdeman is an award-winning strategic brand and design firm. Founded in 1989 by visionary leaders Elle Chute and Denny Gerdeman, Chute Gerdeman has grown into a multi-disciplinary brand experience company with a global presence. With more than 150 awards, including four "Store of the Year" awards, the firm has became one of the most respected design firms in the industry.

Known for creating unforgettable environments, compelling communications and innovative business strategies, Chute Gerdeman connects client, customer and brand. Guided by consumer insights, creative innovation and proven results, Chute delivers unique solutions that best communicate clients' brand and spirit. The firm's core capabilities include intelligence, brand, design and implementation, with expertise across a wide range of retail, restaurant, service provider and CPG brands.

Relationships last when they're built on mutual respect and a foundation of trust, which is why they're honored and privileged when their clients entrust them as partners in their business. Some of their clients have been with them since their inception in 1989, and some are just beginning a relationship that they anticipate will be just as lasting.

Dariel Studio

Dariel Studio is a Shanghai-based international studio specializing in Interior Design. It elaborates unique concepts for living spaces, creative offices, concept stores and high-end hospitality projects (hotels, boutique hotels, SPAS, clubs, bars and restaurants).

The studio was established in 2006 under the name of "Lime 388". Since its creation, the studio has collaborated with luxury brands and international groups such as Christofle, Hermès, MHD, Starwood, Golden Tulip, etc. Dariel Studio also designed the F&B concepts of celebrated chef David Laris (20 venues in China) and has a portfolio of more than 50 projects in Shanghai, Sheshan, Zhouzhuang and Beijing.

2011 was a booming year with several great achievements that were widely published all over the world and that led the Studio to receive the "Best Young Designer of the Year" Award.

Thomas Dariel is the leading designer and founder of the studio. As great grandson of a French furniture designer, grandson of a jazz musician and son of an architect, he has always drawn his strength and inspiration from his family heritage. A Parisian Designer by training and tireless traveler, he came to China in 2006 to launch his own studio, "Lime 388".

After five flourishing years and in the momentum of "Lime 388"'s success, Thomas decided to rename the studio into "Dariel Studio" as a keystone of his artistic maturity and new ambitions.

Design BONO

Design BONO is an interior design group, they consider and present the total organic essence of design through their sensitivity which deviates from five senses (taste, smell, see, touch, hear). Design BONO offers various genres of design such as urban, architecture, residential, interior, product, graphic and fashion design etc.

Design BONO is an authorized company for designs of International corporations, such as Hyundai department store, Shinsegae International (SI), Daemyung Leisure Town, SKII, Doosan Group and so on. In addition, they are working with various projects attaching importance to international luxury brand, such as Chanel and Ballantyne.

Design BONO introduces new design with their creative thinking and execution through visual cooperation with members from all different design fields.

Dear Design & Architecture

Dear Design & Architecture is an interior & industrial design studio based in Barcelona. The team is composed by people from different European countries and fields of competencies. They form a multi-disciplinary team that could be organized in function of each project. Dear creates experiences through conceptual and narrative messages to enhance brands visibility. They are not specialized in any particular market sector, many of their clients come to them for this reason: to reach new fresh ideas and perspectives on their brands.

Dear is the contraction of Design and Architecture. From the beginning, they consider Brand and Product, Design and Architecture are a great whole as they are inextricably tied. They believe in little pleasures and beautiful things, functionality and quality. Dear's goal is to be a reference for their partners and clients.

Duccio Grassi Architects

Duccio Grassi Architects develops projects and concept designs for buildings, boutiques and showrooms for Max Mara, Zara, Canali, Guru, Guess by Marciano, Penny Black, Max&Co, Ceramiche Refin, Iris Ceramica, among the other brands. DGA also designs private villas, restaurants, jewelries, malls and offices, among which the extension of the Wafi City Mall in Dubai, the Al Hamra Mall, the Al Yasra Headquarters and the Mazaya Headquarters in Kuwait. DGA has participated to various Salone del Mobile in Milan designing pieces of furniture for Emmebi, Minotti Cucine and ViaBizzuno.

DGA's projects received awards in Paris, New York and Los Angeles. DGA has offices in Milan and Reggio Emilia.

Fabio Novembre

Fabio Novembre was born in Lecce in 1966. In 1984 he moved to Milan where he graduated in Architecture at Politecnico. In 1992 he lived in New York where he attended a Cinema course at the New York University.

During his American stay he got to know Anna Molinari and he realized for her his first interior project: the shop "Anna Molinari Blumarine" in Hong Kong. In the same year he opened his studio in Milan.

The collaborations with leading design companies intensify during the years, Cappellini, Driade, Meritalia, Flaminia and Casamania just for naming the main important ones; at the same time the showroom projects and boutique for the best international fashion brands going on as the Tardini shop in New York, the Blumarine store in London, Singapore and Taipei as well as the Meltin' pot and the Stuart Weitzman shops all around the world, from Rome to Beijing.

In 2008 the Comune of Milan dedicates a solo exhibition in the Rotonda di Via Besana as prestigiuos location named " Teach me the freedom of swallows", while in 2009 the Triennale Design Museum of Milan invited him to create a personal exhibition named "Il Fiore di Novembre". In 2010 the Comune of Milan charges him of an exhibition inside the Italian Pavillion on the occasion of the Shanghai Expo.

2011 is the photography year: after art-directing of the exhibition "Lavazza con te partirò" at the Teatro dell'Arte of the Triennale of Milan on the occasion of the 20th anniversary of the company's calendar; he also designed and curated the Steve McCurry exhibition at MACRO Testaccio, Rome.

In April 2012 he signs the new exhibition setting for the fifth edition of Triennale Design Museum.

Four-by-Two

Four-by-Two has offices in London, Edinburgh and Muscat, Oman. Formed in 2002, Four-by-Two provides concept design and roll-out design management services to the retail sector and to the drinks and leisure industries. A balance of creative thinking and practical knowledge means that project work combines the best of aspirational and practical design – creating environments that are engaging for the customer, yet operationally appropriate, with flexibility and manageability for the client.

Guan Interior Design Co.,Ltd.

Guan Interior Design Co.,Ltd. was established in 2007. Specialized in interior design of commercial space but not being confined to a sole industry, the firm has its influence into the fields of clothing, real estates, entertainment and catering.

Guan Interior Design keeps an eye on the trends of various aesthetic elements of fashionable, classic and modern, sticks to the environmental friendly design concept and always seeks for the breakthrough and innovation.

Jaklitsch / Gardner Architects PC

Jaklitsch / Gardner Architects (J/GA) is an award-winning, New York City-based studio with an expertise in designing high-end commercial and residential buildings and interiors, furnishing and objects. Over the firm's 15 year history, J/GA has built several-hundred projects throughout North and South America, Europe, Asia and the Middle East – giving the firm an international perspective on design and project management. J/GA has collaborated with fashion brands including Marc Jacobs International, for whom J/GA has designed all retail concepts for over 13 years, as well as Shelly Steffee, Moscot Eyewear and Shinsegae International. J/GA's work has been exhibited widely and featured in numerous publications, including Architectural Digest, Elle Décor, The New York Times, Time Magazine, Wall Street Journal, Azure, Hinge, Surface, and Wallpaper.

The firm's principals, Stephan Jaklitsch and Mark Gardner, are actively involved in all stages of every project and operate under the principle that architecture and the built environment posses the ability to communicate collective values, provide relevance and meaning through the experience of a place. They take the lead with research and investigations to develop an appropriate response to the unique needs of each project with sensitivity to context, materiality, form, space, light and sustainability goals.

Joanna Laajisto Creative Studio

Joanna Laajisto Creative Studio is a Helsinki based design company which main focus is creating commercial interiors and brand identities for retail and hospitality field.

Joanna Laajisto, the founder of Joanna Laajisto Creative Studio, is a Helsinki based interior architect and designer who has studied and made her career in the West Coast of United States. Prior to moving back to her native country Finland, she worked in Los Angeles at an international architect firm designing large scale commercial projects.

In addition to her BA degree in interior architecture, she is also a LEED accredited designer, making her an expert in environmental and energy efficient design.

The work of Joanna Laajisto Creative Studio is simultaneously driven by functionality and aesthetics. The philosophy is not to clutter this world with unnecessary things but to find the hidden beauty of each space and to enhance it by creative solutions. The company believes that a successful design is a 360 degree experience. To achieve the most wholesome end results possible, they create projects from conceptual ideas to service design, graphics, copy,

Kengo Kuma & Associates

Kengo Kuma was born in 1954. He completed his master's degree at the University of Tokyo in 1979. After studying at Columbia University as Visiting Scholar, he established Kengo Kuma & Associates 1990. In 2009, he was installed as Professor at the Graduate School of Architecture, University of Tokyo.

Among Kuma's major works are Kirosan Observatory (1995), Water/Glass (1995, received AIA Benedictus Award), Stage in Forest, Toyoma Center for Performance Arts (received 1997 Architectural Institute of Japan Annual Award), Bato-machi Hiroshige Museum (received The Murano Prize). Recent works include Nezu Museum (2009, Tokyo), Yusuhara Marche and Wooden Bridge Museum (2010), Asakusa Culture and Tourism Center (2012) and Nagaoka City Hall Aore. (2012) A number of large projects are also going on abroad, including the arts centre in Besancon City, France, and a new Victoria & Albert Museum building in Dundee, Scotland U.K..

Iosa Ghini Associates

Iosa Ghini Associates has an office in Bologna in an ancient palace and another in Milan, where architects, engineers and designers of different nationalities work together.

Founded in 1990 during the time it acquired a particular expertise in developing projects for large groups and developers that operate internationally.

The professional development of the society grows in the design of commercial and museum architectural spaces, planning areas and structures dedicated to public transport and design of worldwide retail chain stores.

Among the most recent major projects are the Ferrari Store in Europe, USA and Asia; multifunctional residential project in Budapest, several hotels in Europe (including Budapest, Nice and Bari) and the airport areas of the airline Alitalia.

Furthermore, the metro station Kröpcke Hannover (Germany 2000), The Collection shopping center in Miami, USA, (2002), the Galleria Ferrari Museum in Maranello, Modena (2004), the headquarters of Seat Pagine Gialle, Turin (2009), and finally the Casa Museo Giorgio Morandi (2009), the design of transport infrastructure People Mover in Bologna (2010), and the latest IBM Software Executive Briefing Center, Rome (2010).

Makoto Tanijiri / Suppose Design Office

Makoto Tanijiri born in Hiroshima in 1974, is an architect and established suppose design office in 2000.

He has designed over 60 residences and his work sphere is broad including commercial spaces, set-ups of exhibition spaces and products.

Maurice Mentjens Design

Maurice Mentjens primarily designs interiors and related objects and furnishings. Creations are almost exclusively for the project sector: shops, hotels and restaurants, offices and museums. The aim of this compact, talented and dynamic team is to deliver high-end design reflecting its passion.

Quality and creativity are prioritised in all aspects of the design and implementation process. The agency is three times winner of the Dutch Design Award.

In 2007, the design agency received the Design Award of the Federal Republic of Germany. Clients include the Bonnefantenmuseum Maastricht, Frans Hals Museum Haarlem, Post Panic video producers Amsterdam, DSM in Heerlen and Amsterdam Airport Schiphol.

Miralles Tagliabue EMBT

Founded in 1994 by Benedetta Tagliabue and Enric Miralles, Miralles Tagliabue EMBT has a main office in Barcelona and a branch in Shanghai, operating all over the world. Miralles Tagliabue EMBT can be seen as a melting pot of ideas, a meeting point of traditions and innovations; where each project presents a challenge and at the same time a new learning opportunity.

It's without question an open approach, full of exploration and experimenting. Even so the high level of conceptual thought is fundamental. The studio itself reflects the belief of changing the environment by observing and respecting the site, its history and culture.

EMBT's work includes various iconic buildings and public spaces to the city of Barcelona: Gas Natural tower, Santa Caterina Market, Diagonal Mar Park. It also includes a number of high profile projects in other European cities Scottish Parliament in Edinburgh, Utrecht's Town Council and Music School in Hamburg.

Today under the direction of Benedetta, EMBT works not only within the scope of architecture, but also within that of landscape, urban infrastructure, rehabilitation and design, trying to preserve the spirit of the Spanish and Italian artisan architectural studio tradition.

MOMENT Inc.

Hisaaki Hirawata and Tomohiro Watabe established MOMENT Inc. in 2005. Their projects are wide-ranging as graphic, product, interior, and architecture. Hirawata and Watabe attempt to suggest strong design-message which is not compressed in a category. 2D and 3D is an equal theme to every kinds of design project.

NC Design & Architecture Ltd.

NC Design & Architecture Ltd. (NCDA) is a young Hong Kong based multidisciplinary design practice that believes on an all encompassing approach to design. It integrates seamlessly between architecture, interior, furniture, product, uniform and graphic design, extending an aesthetic from the mood of a space right down to its detail components. Each project is approached without a pre-determined style, but through intense research and close communication with clients, consultants & specialists, deriving solutions with a strong concept that not only meets the client's functional & budget requirements, but also creates a memorable & inspiring experience.

Nelson Chow is a New York State licensed Architect and the founder for NCDA. After receiving his Master of Architecture from The University of Waterloo in 2004 and Menswear Tailoring Certificate at the Fashion Institute of Technology in New York City in 2005, Nelson worked for design and concept firm AvroKO in New York City on a number of critically acclaimed projects such as Steelworks Lofts, W hotel and Madame Geneva Bar. Upon receiving his New York State Architectural license, Nelson took a position at the award winning Hong Kong practice Edge Design Institute designing privately commissioned houses and yachts, as well as commercial serviced apartments.

NCDA is currently working on a number of highly anticipated projects, including 2 fashion boutiques, 3 Residences and 3 Restaurants in Hong Kong.

OMA

OMA is a leading international partnership practicing architecture, urbanism, and cultural analysis. OMA was founded in 1975 by Rem Koolhaas as a collaborative office practicing architecture and urbanism. OMA is led by seven partners – Rem Koolhaas, Ellen van Loon, Reinier de Graaf, Shohei Shigematsu, Iyad Alsaka, David Gianotten and Managing Partner, Victor van der Chijs – and sustains an international practice with offices in Rotterdam, New York, Beijing, Hong Kong, and soon Doha.

OMA's New York office was established in 2001. Under the direction of partner Shohei Shigematsu, the New York office has most recently overseen the completion of Milstein Hall at Cornell University as well as a seven screen pavilion in Cannes for Kanye West. OMA New York is also overseeing the construction of the Musée national des beaux-arts du Quebec and an artist studio in New York city, while developing designs for the Marina Abramovic Institute in upstate New York, a cultural center in Miami, and an art foundation in Manila, Philippines.

OOBIQ Architects

OOBIQ Architects is an award-winning architectural and design firm based in Hong Kong and Singapore. OOBIQ Architects' founding director is Italian architect Samuele Martelli, graduate of the University of Florence where he also have lectured, researched and supervised undergraduates of architecture.

After years of experience in European design firms working on projects all around the world, and after experienced in Hong Kong and Shanghai, Mr Martelli launched his own design firm, to express his notable architectural skills, and bring Italian quality and style to Asia.

OOBIQ Architects' design services include architecture, interior design, and event and product design, artistic direction and project management. Oobiq Architects are positioned to manage diverse projects with superior creativity and knowledge.

OOBIQ Architects' clients are both European and Chinese brands with high visibility who demand a quality product with personalized service. Mr Martelli, often invited to conferences and seminars to share with guests his experience and design philosophy, was also awarded the '40 under 40 Design Stars of Tomorrow' by Perspective Magazine, Hong Kong, in 2012.

Pedrocchi Architects

Pedrocchi Architects was established 2011 by Reto Pedrocchi in Basel. He studied in Basel and Berlin. After his studies he worked for Herzog & de Meuron Architects in Basel and Tokyo and was the person in charge for the realization of the Flagship store of PRADA in Tokyo. After the collaboration at Herzog & de Meuron Architects he was teaching assistant and head of teaching for Professor Christian Kerez at the ETH Zurich from 2003 to 2009.

From 2005 he carried on his own business together with a partner. 2011 he decided to compile architecture exclusively under his own name and founded his office, Pedrocchi Architects at Basel. The office history is still recent, although Reto Pedrocchi could receive recently already the price for the Alpine Interior Award 2011. Since 2011, he also continues his teaching activity as a visiting Professor at the HSLU, University of Luzern.

plajer & franz studio

plajer & franz studio is an international and interdisciplinary team of 50 architects, interior architects and graphic designers based in Berlin . All project stages – from concept to design as well as roll-out supervision – are carried out in-house. Special project-based teams work on overall interior and building construction projects and on communication and graphic design.

The company's client list includes karl lagerfeld, galeries lafayette, s.oliver, bmw, mini, Puma, timberland, pierre cardin, and salewa. plajer & franz studio has also established itself in the premium sector of luxury residential projects and hotels in both Europe and Asia; these include a recently completed hotel in Porto, a five-star resort in Croatia and 50,000 m^2 premium apartments on the Portuguese coast, both in development. Several other major projects, for example luxurious villas in Thailand and Kazakhstan are also currently in the design phase.

From private yacht to automobile trade stand via award-winning bars and luxury apartments – the key to plajer & franz studio's freshness of vision lies in their continuous exploration and cross-fertilisation between disciplines and areas of experience. Their ability to deliver show-stoppingly innovative design with elegant and meticulous finishing down to the smallest detail lies in being able to take what they learn in one area and applying it, where appropriate, in another: high tech material forming from the car industry, for example, may yield exciting new surfaces for a shop-in-shop project whereas new developments in the use of digital display techniques from the bar and club scene might fit perfectly with a new automobile display concept – it's all in the mix!

Quadrangle Architects

Established in 1986, Quadrangle Architects Limited is among Canada's most dynamic architectural firms.

Quadrangle's portfolio of projects illustrates a diversified client list that includes major players in the residential, commercial, media and hospitality industries. Equally wide-ranging is their scope and expertise in the rehabilitation and conversion of existing buildings, historical restorations and renovations, corporate interiors and institutional projects. With over 50 awards to their credit, they consistently achieve excellence in architecture and design.

Led by principals Brian Curtner, Jeff Hardy, Les Klein, Sheldon Levitt, Anna Madeira, Caroline Robbie, Susan Ruptash, Ted Shore and Richard Witt, Quadrangle is made up of a talented 100 person multi-disciplinary team. They pride themselves on a hand on approach where each project is led by a principal and project manager, providing continuity and a single point of contact for the client throughout the project.

Inspired by the belief that architecture should be approached as creative problem-solving, they base success on strong long-term relationships with clients. This close association allows the Quadrangle team to grasp and effectively respond to a client's requirements, expectations and budget, and helps them address site and user needs sensitively and responsibly.

The methodology of architecture and design Quadrangle employ is also very much guided by relationships. They listen to a client's needs and base the creative end product on each individual project's culture, context and requirements. The diversity that results leads to operationally successful buildings, visually engaging environments and stimulating urban developments.

Recent projects of note include The Address at High Park, West Harbour City, Studio on Richmond, 130 Bloor Street West/155 Cumberland Street, the BMW Toronto Showroom at the bottom of the Don Valley Parkway and the new ONroute Service Centres on Highways 401 and 400. Recent interiors projects include Corus Quay, the Deaf Culture Centre in the Distillery District, the Toronto offices of Cossette Communication Group and Citytv and OMNI Television at Yonge-Dundas Square.

SAQ Architects

SAQ Architects is a conceptual and interdisciplinary design agency specialized in developing spatial sceneries and concepts. The practice is formed through a studio environment where architects, interior designers, urbanists, video artists and graphic designers team up according to the specific orientation or necessities of each project. SAQ believes strongly in cooperation and regularly invites professional experts or specialists to participate in the materialization or the elaboration of an idea.

Each project undertaken is a reason to broaden perspectives. The subjects on which SAQ has been asked to focus many: the studio has contributed concepts for marketing strategies as well as for temporary installations in public events all within the constantly evolving framework of societal and technological evolution.

In this versatility, every project or concept undertaken by SAQ aspires to deliver to its end-user powerful and moving experiences where all senses can be stimulated and where the materialized space forms an event in itself.

SAQ considers the relationship with the client as a vital element in the design process. A successful project is always the fruit of passionate and constructive debates between client and studio with the ultimate goal the satisfaction of the end user. Acclaimed projects such as the fashion concept store L ECLAIREUR (Paris-FR), or the restaurant KWINT (Brussels-BE) are some of the latest high-end realizations of SAQ's interior architecture department. Currently, SAQ is also engaged in the masterplan outline for mixed use development in Brussels and Berlin where the offices are equally located.

SAKO Architects

SAKO Architects is a Beijing based architecture firm of Japanese architect Keiichiro Sako. So far SAKO Architects has produced over 70 projects in China, Japan, Korea, Mongolia and Spain. In addition to the core services of architectural design and interior design, they also provide graphic design, furniture and urban master planning. Nowadays, they continue to develop the architecture known as, "CHINESE BRAND ARCHITECTS," using specific themes which can only be found in China.

Sid Lee Architecture

Sid Lee Architecture is an architectural and urban creativity firm, based in Montréal (Canada), with a satellite office in Amsterdam (Holland). Founded in 2009 with the integration of the architectural firm Nomade (founded in 1999), Sid Lee Architecture is the result of the combined talents and skills of architects and urban designers Jean Pelland and Martin Leblanc, as well as Sid Lee, a well established creative agency. Sid Lee Architecture is a creative endeavour, exploring urban design and architecture through an innovative combination of collaborative design processes, media and technology applications. Challenging conventionality, this broad approach generates a new kind of design knowledge and cross-disciplinary culture, empowering the firm to work on a large diversity of urban and architectural ventures, from interior design to urban planning. This new design culture succeeds in delivering landmark projects of diverse sizes that embody the demands and expectations of today's clients. The firm has distinguished itself not only by its design excellence but also by high client fidelity.

sinato Inc.

sinato Established by Chikara Ohno in 2004, their service include architectural design, interior design and the like.

Chikara Ohno, born in Osaka, Japan in 1976. In 1999, he graduated from department of civil engineering, Kanazawa University. In 2004, he established sinato Inc. Chikara became part-time lecturer of Kyoto University of Art and Design.

The firm received various domestic and international awards, such as IIDA Global Excellence Awards, Honorable Mention, The Ring iC@ward international design 2010, Gold Award, Contractworld.award, New Generation Short-listed, Design For Asia, Bronze Award, FUJITSU DESIGN AWARD 2011, judge's special award, Good Design Award, Winner, JCD Design Award, Gold Award, SDA Awards, Grand Prix, Display Design Award, Encouragement Prize, NDF Awards, Encouragement Prize, Best Store Of The Year, Excellent Prize, Mie Architecture Awards, Winner, Gunma Agricultural Technology Center Design Competition, Honorable Mention, TOTO Remodel Style Contest, Grand Prix, NASHOP Lighting Awards, Excellent Prize and more.

Silvia Simionato Architetto

Silvia Simionato loves the materials, surfaces and lighting; these materials allow it to create emotional projects. Modern art also contributes to the inspiration of their projects.

Space and its transformation is the subject of the project, each project must create an emotion to those who live the place.

Silvia Simionato has developed an international reputation for continuously innovative.

Design concepts in many areas from architecture, retail and restaurant design to interior architecture, SPA and wellness, nautical, furniture and graphic design.

Silvia Simionato born in 1967, lives and works near Venice, Italy. Member CNA Architect of Venice - Italy.

Specialnormal Inc.

Specialnormal Inc. is a multidisciplinary design office based in the heart of Tokyo. It was founded by Shin Takahashi in 2011, and within a year it already built a great portfolio such as Note et Silence, NIKEBASKETBALL, Quiksilver Store Toyosu and Baccarat 2011 S/S Collection.

Shin Takahashi is a founder and principal designer of Specialnormal Inc. He studied an interior design at Kuwasawa Design Institute. Prior to founding Specialnormal Inc., he was engaged in several projects with world-class retail brands.

Storeage

Storeage is an Amsterdam-based retail strategy and design agency. The agency's goal is to redefine the retail experience for its clients and their customers, based on its deep understanding of brands and consumers.

Storeage views retail as a platform where product and consumer meet, where stories are shared and interaction is key. Where shopping becomes branding and each experience makes the difference. Its designs are not just about selling products - they allow consumers to develop a lasting relationship with a brand and to experience it with all their senses.

Storeage's creative team is made up of the best minds and eyes in the industry. This close-knit group of retail architects, communication experts, interactive designers, visual merchandisers, marketing and branding professionals incorporates ten nationalities with studios in Europe and Asia. Sharing differences gives Storeage a global perspective, keeps things fresh and allows it to develop a market specific approach for each retail challenge.

Stefano Tordiglione Design Ltd.

Stefano Tordiglione Design Ltd. was founded by Mr Stefano Tordiglione in Hong Kong. His art direction is characterized by East meeting West that every detail is a result of research and intuition. With the "Mind in Italy", Stefano strives to integrate Italian design while preserving local cultures.

Stefano Tordiglione Design provides multi-disciplinary professional interior design and architectural services to exclusive clients in the following areas: Hospitality, Retail, Office, Residential, Architecture, Urban planning, Product, Graphic + Packaging, Art direction and design consultancy service.

Stefano was born in Napoli, Italy and he studied and worked in New York and London for over ten years. His first design experience started in London in 1991. Since then he has worked for internationally acclaimed Italian studios specialized in retail, luxury hotel and resorts, residential development and private yachts.

He is also an artist and his works are part of international private contemporary art collections. It's also worthy to mention the fact that he worked for UNICEF as Project and Art Director in several international events organized in Italy. His numerous artistic experiences have made him develop excellent intuition into aesthetic and design.

The design team at Stefano Tordiglione Design Ltd has strong international background with architects and designer who have worked and studied in Italy, USA, Australia and Korea.

The main office is based in Hong Kong. It also works in partnership with architectural and interior design studios in Italy, including Rome, Milan, Florence and Bari.

Studio 63 Architecture + Design

Studio 63 Architecture + Design is based in the historical center of Florence Italy. The fruitful encounter between Piero Angelo Orecchioni and Massimo Dei led to the foundation of Studio 63 in 1998. By 2003, Studio 63 inaugurated its New York City office, by 2005 its office in Hong Kong and by 2008 its office in Shanghai and an operative partnership in Dubai and Singapore.

The creative team is composed of gifted professionals, coming from various disciplines, working together in a fertile and challenging multi-cultural exchange.

A strong identity is the hallmark of our projects. This identity is the result of extended research, creative proposals and deep respect of the contemporary language criteria.

Studio 63 is operating in more than 25 countries around the world managing to the last detail projects from concept development through design and planning towards their complete achievement. Our specialties are retail design and hotelier.

The creative process is an intense journey shared with the client. Only close collaboration and understanding can lead a project to new horizons. This practice is inspired by a continuous research and the blend of various artistic disciplines. Each project is a new opportunity for growth and development.

ARTPOWER

About ARTPOWER

PLANNING OF PUBLISHING

Independent plan, solicit contribution, printing, sales of books covering architecture, interior, graphic, landscape and property development.

BOOK DISTRIBUTION

Publishing and acting agency for various art design books. We support in-city call order, door to door service, mail and online order etc.

COPYRIGHT COOPERATION

To further expand international cooperation, enrichpublication varieties and meet readers' multi-level needs, we stick to seeking and pioneering spirit all the way and positively seek copyright trade cooperation with excellent publishing organizations both at home and abroad.

PORTFOLIO

We can edit and publish magazine/portfolio for enterprises or design studios according to their needs.

BOOKS OF PROPERTY DEVELOPMENT AND OPERATION

We organize the publication of books about property development, providing models of property project planning and operation management for real estate developer, real estate consulting company, etc.

Introduction OF ACS MAGAZINE

ACS is a professional bimonthly magazine specializing in high-end space design. It is color printing, with 168 pages and the size of 245*325mm. There are six issues which are released in the even months every year. Featured in both Chinese and English, ACS is distributed nationwide and overseas. As the most cutting-edge counseling magazine, ACS provides readers with the latest works of the very best architects and interior designers and leads the new fashion in space design. "Present the best whole-heartedly, with books as media" is always our slogan. ACS will be dedicated to building the bridge between art and design and creating the platform for within-industry communication.

ARTPOWER

Artpower International Publishing Co., Ltd.

Add: G009, Floor 7th, Yimao Centre, Meiyuan Road, Luohu District, Shenzhen, China
Contact: Ms. Wang
Tel: +86 755 8291 3355
Web: www.artpower.com.cn
E-mail: rainly@artpower.com.cn

QR (Quick Response) Code of ACS Official Wechat Account

Acknowledgements

We would like to thank all the designers and companies who made significant contributions to the compilation of this book. Without them, this project would not have been possible. We would also like to thank many others whose names did not appear on the credits, but made specific input and support for the project from beginning to end.

Future Editions

If you would like to contribute to the next edition of Artpower, please email us your details to: artpower@artpower.com.cn